Toward
a New Foundation
for Physics

Toward
a New Foundation
for Physics

Joseph H. Spigelman

Kips Bay Press
New York

Copyright © 1987 by Joseph H. Spigelman

Published by
Kips Bay Press
333 East 30th Street
New York, New York 10016

Library of Congress Catalog Card Number 87-82142

ISBN 0-9619183-0-6

Printed in the United States of America

Contents

Preface

"Our experience hitherto justifies us in believing that nature is the realization of the simplest conceivable mathematical ideas." A. Einstein: Lecture delivered at Oxford University, June 10, 1933.

". . . l'idée de mon oeuvre était dans ma tête, toujours la même, en perpétual devenir. Mais elle aussi m'était devenue importune. . . . Chez moi les forces de l'écrivain n'était plus a la hauteur des exigenes égoistes de l'oeuvre." M. Proust: *Le Temps Retrouvé*

"I can't go on. I must go on. I'll go on." S. Beckett

"You don't have to finish the work, but you're not allowed to stop trying." *Ethics of the Fathers*

". . . . it was a great mistake in my youthful studies that I made the design of the building too great. The result was that I could not finish the upper story, indeed could not even put on the roof." G. C. Lichtenberg

". . . meine Kraft zur Bewaltingung der Aufgabe zu Gering ist. . . . Mögen andere kommen und es besser machen." L. Wittgenstein: *Tractatus Logico-Philosophicus*

Some personal history may prepare the reader for the work that follows.

The work originated in the spring of 1935. I was twenty years old then, an English teacher at the City College of New York (from which I had graduated the preceding January) and a post-graduate student at Columbia University. Among my courses was one in esthetics. The professor read his lecture in a monotonous drone that put me, one balmy day, to sleep. I awakened to hear him mumbling something about "ineffable beauty" and "empathic harmony." Thinking confusedly that I had wandered into a class in mystical theology, I picked up my books and papers and left the class, never again to return to formal study. I then embarked on a Cartesian quest, from esthetics to psychology, from there through biology, chemistry, and physics to mathematics and its foundations, trying to find something at once clear and unquestionably true. (Though I had "majored" in English and "minored" in philosophy as an undergraduate, my B.S. matriculation required that I take a good deal of course work in the natural sciences and in mathematics, so that I could grope my way, however incompetently, around these fields. That I was teaching only two evenings a week and had my summers and long academic holidays free was also helpful.) After about a year of such searching, all I had come up with was a feeling that I understood the meaning of the integers — that 2 means one and another; 3, one and another and still another, etc. I recognized that this understanding was trivial and otiose and I began thinking seriously about abandoning my quest and returning to studies in English in order to salvage my career (in jeopardy because of my neglect) and to advance in it.

Before I could firm this decision, I quarreled with my girl-friend at the time. Immediately after the quarrel (in her apartment overlooking New York's Central Park), I rushed into the park in a state of extreme agitation and spent the whole night there alone. After venting my rage at my friend, I experienced a moment of illumination, in which I conceived a way of formulating an object language, based on my definition of the whole numbers, that could be used to construct models of natural structures and processes that could serve to make them intelligible. During the next year and a half, I worked feverishly, hardly sleeping, neglecting my teaching even more scandalously than theretofore, to develop my formalism and to build my models, producing in the process more than a thousand pages of typescript. (My state of mind during that period is well described by A. Camus, in *The Myth of Sisyphus*: "If thought discovered in the shimmering mirrors of phenomena eternal relations capable of summing them up and summing themselves up in a single principle, then would be seen an intellectual joy of which the myth of the blessed would be but a ridiculous imitation," and by G. Steiner (in *A Reader*): "The absolute scholar is a rather uncanny being. He is instinct with Nietzsche's finding that to be interested in something, to be totally interested in it, is a libidinal thrust more powerful than love or hatred, more tenacious than faith or friendship, not infrequently, indeed, more compelling than personal life itself.")

That state of mind could not endure. My "joy," my "libidinal thrust," were undermined by physical exhaustion, by the inadequacy of my income even for bare subsistence, by indications that I faced imminent dismissal from my teaching position unless I "shaped up" to it, above all, by my recognition that my powers were unequal to the task I had so romantically undertaken. In 1938, after another night of soul-searching in Central Park, I abandoned the work. (I continued teaching for a while, then became successively a writer with *Fortune* Magazine, a planner of economic mobilization for war and demobilization for peace with the U.S. War Production Board, a freelance writer on economic and political subjects, a longshoreman, an executive of a trucking company, an encyclopedia editor, and from the spring of 1960 to my retirement at the end of 1981, director of research for a company concerned with investment counseling, portfolio management and business consulting.)

During the years since 1938, friends and associates to greater or lesser degree acquainted with my work through conversations with me, attendance at informal seminars I conducted in the late 1930s and late 1940s, and reading my manuscripts, argued (with whatever justification) that there was "something" that should not be allowed to die with me. I resisted their urgings, for many compelling reasons:

• I was utterly an outsider, unknown to the scientific community and with no academic standing (though my publications in politics, economics, and finance and my business consulting had gained me a certain reputation in non-academic circles). I could not regard myself, or be fairly regarded, as scientist, mathematician, philosopher, or scholar. My work didn't bear the stigmata of discourse in physics. It was not thickly studded with mathematical equations. It made no quantitative predictions that might be subject to testing. It didn't add to the mass of empirical data nor refine the values of physical constants to more decimal places. It didn't directly participate in the discussions and controversies then agitating the world of physics. I didn't see how I might contribute to these discussions nor make a difference to them. For these reasons, qualified people to whom I might send accounts of my work, or who might chance upon it should it achieve printed form, could hardly be expected to give it the painstaking reading required for comprehension and proper judgment.

• Particularly so, because of scientists' characteristic conservatism and exclusivity, especially in physics, the "hardest" of the sciences. They are strongly inclined to be intolerant and disdainful toward those who haven't endured the rigorous discipline required to gain Ph.D.s and to secure academic appointments. They resist new ideas, especially from outsiders — and even Einstein was considered an outsider before 1905 — unless they prove indispensable. [Typifying their attitude is a comment by H. Pagels (in *The Cosmic Code,* New York: Simon & Schuster, 1982): "Physicists are conservative. . . . [In contrast,] Pseudoscientists lack that commitment to existing principles, preferring instead to introduce all sorts of ideas from the outside."] This resistance to the ideas of outsiders is understandable, for were physicists — and especially the more prominent among them — to pay attention to all such ideas with which they might be presented, they would hardly have time left for reckoning with the ideas of their established colleagues or to do their own work. (As S. Hawking remarked in his Inaugural Lecture as Lucasian Professor of Mathematics at Cambridge University: "I get two or three 'unified theories' in the mail every week.")

• Most important, though knowing full well that I lacked the knowledge of physics and the command of mathematics needed to fulfill my work, I lacked the strength, the perseverance, and, for long periods when I was fully preoccupied with earning my livelihood, the time needed to remedy these deficiencies. I came to know a little more about physics than I did in the mid-1930's, but the passage of years weakened my powers of concentrated thought, sapped my confidence, and cooled the ardor for work I exhibited when I was young. Unable to cope with the technical literature in physics — with which even professional physicists are less and less able to cope, except possibly in their own increasingly narrow specialities — I have had to depend on the writings for laymen of such eminent physicists as Einstein, Bohr, Heisenberg, Schrödinger, and Feynman, among others, on works in the philosophy of science, on semi-popular and even blatantly popular publications, though recognizing their superficiality, their misleading translations of mathematical abstractions into language accessible to the uninitiated, often their glibness and pretentiousness.

Despite these deterrents, my employer prevailed upon me some ten years ago to prepare some intimation of the work of my youth for publication — granting me a half-year's leave of absence at full-pay to enable me to do so. From the renewed effort, there eventually emerged a paper published in the October 1978 issue of *Foundations of Physics*. The paper elicited a few expressions of interest and encouragement — and then silence. Nevertheless, on my seventieth birthday, in October 1984, I decided on one last effort to deliver my work, to correct, clarify, and elaborate the argument of my article and, if possible, to get the outcome published as a book.

My reasons for this decision:

• The only time I can, with some justification, be said to have been truly creative was that moment a half century ago in Central Park. Only the work proceeding from that moment could possibly have lasting significance.

• I need to clear my conscience, to rid myself of my obsession, to exorcise it, delivering back what was once "given" to me — though it be a still-birth or a monster, though it make me look foolish or deluded, though it fail utterly to impress the community of physicists or anybody else who matters, though I am now too old and tired to be able to get much joy of any recognition it may conceivably achieve; yet must I deliver it. I must free myself of my work by being shown that it is vitiated by a fatal logical flaw or that it is irrelevant to the concern of physicists and cannot reasonably be expected to be made relevant; or by somehow persuading others more capable than myself of developing the work and fulfilling its intentions to do so.

• I am convinced, as Einstein was, that "most of the essential ideas of sciences are basically simple and may, as a rule, be expressed in a language comprehensible to everyone," that only a truly simple theory is worth serious consideration, and that the ultimate test of a theory is "the beauty of its equations," which depend on their simplicity, generality, and clarity.

• I feel that, however unlikely, it is still possible that an outsider may come up with some idea of value to the scientific establishment. ["It is also possible that the correct picture will emerge from some totally new idea. In the words of Niels Bohr, it may be that our present ideas 'are not sufficiently crazy to be correct'" (H. Harari in *Scientific American*, April 1983).]

So much for *my* reasons for writing and publishing this book. But now, one may well ask, what reasons might anyone else have for reading it, let alone for giving it the close attention required for its comprehension and evaluation. Essentially, just one reason: the same feeling I have just ascribed to myself: the feeling that, despite the odds, there may be "something there," that it may truly be significant, and that the *active* reader may be able to put it to good use.*

What is attempted here is not a definitive treatise but an essay intended to incite work by others. Because I am not a scholar and would not pretend to be one, I shall not attempt to make the work appear scholarly. There is no index, no bibliography. Citations, though (I feel) almost always sufficient to guide the interested reader to the work cited, will not be given in full (with volume and page number, etc.) as demanded by examiners of Ph.D. candidates and editors of scientific journals.

There will be no rehashing of academic disputes, no

---

*I have underscored "active" because serious reading is always active, not passive as it appears to be. The reader, unlike the listener to a lecture or the watcher of a demonstration, is in control. He (or she) can read rapidly or slowly and he can vary his pace at will; he can pause to ponder and question what he has just read; he can skim or skip what appears redundant, irrelevant, or unimportant; he can drop the book, magazine, or whatever at any point and return to it when he is "in the mood" for it; he can return to a passage that he didn't understand at first try, or that he may have accepted too uncritically, for a clearer and more critical understanding; he can correct errors, whether of typography, style, fact, or concept; he can go back and forth in the text, and back and forth again and again, checking for consistency, making connections that the author didn't see, interpolating ideas the author may not have conceived or facts he may not have known (but that the reader feels may be pertinent), transforming the presentation and the matter presented in any way he may choose. In a word, he can *appropriate* what he reads much more surely and thoroughly than the listener or watcher can. Only dialogue can rival reading as a means of informing the mind. But the participant runs the risk of being cowed into passivity or acquiescence by the prestige, confidence, or self-assertiveness of his interlocutor. The reader is exposed to similar risk, but to a much lesser degree.

mathematical or graphical embellishment, in the manner of textbooks and papers in the learned journals. Should a writer have made a certain point more elegantly or forcefully than I could have done, I quote him, not for support of my argument but for esthetic enhancement of it. If I feel that a particular citation may help the reader place my argument in its intellectual context, I shall note the citation. Since I have no academic standing, I cannot speak with intrinsic or imputed authority. As with a work issued anonymously, my argument must stand by itself, speak with whatever authority it can itself command. Standing naked, without concealing scholarly paraphernalia, the work's validity and possible significance can be more readily determined.

Finally, it should be emphasized that what I present here is not a theory, rivalling those currently accepted or seriously considered in physics, but a conceptual framework intended to make such theories intelligible. And even for this limited purpose, it can only point the way for others, better qualified for the task, to pursue. Like Moses on Mount Pisgah, I have seen what I take to be the Promised Land, but I cannot go there. Let others decide whether that land is worth entering, exploring, and developing.

I am grateful for the encouragement received from friends who are not themselves physicists or philosophers of science, especially from Dr. Leo Bogart, Richard Laupot, Erika Engels, Dr. Alfred Winslow Jones, Richard Storrs Childs, Immanuel and Elsie Elston, Dr. J. Frederick Dewhurst; and from my family. I am grateful for the criticism that drafts of the paper from which this work germinated received from Professor Sidney Morgenbesser and Professor James Higginbottom of Columbia University, Professor Robert MacNab of Yale University, Dr. Aage Peterson of Yeshiva University, Professor David Finkelstein of Georgia Tech, and Professor John Archibald Wheeler of the University of Texas. My greatest debts are to Dr. David Elkin, who helped clarify my thinking and to give it more precise and acceptable expression, and to the late Dr. Julian Gumperz, the employer who encouraged me to return to the work of my youth and who gave me the financial support that enabled me to do so.

I bear sole responsibility for the views expressed in this book.

Chapter 1
The Problem
of Intelligibility

The twentieth century has witnessed the greatest achievements in the history of physics — but also its greatest crisis.

First, the extraordinary empirical findings:

1. Using the new technology of radio, infrared, X-ray and gamma-ray, as well as optical astronomy, satellite-mounted telescopes, digitalized and computerized spectral analysis and image enhancement, etc., astronomers and astrophysicists have discerned billions of galaxies, each containing tens of millions or even billions of stars, and their grouping in clusters and super-clusters of galaxies; the evidence of stars being born and dying; such starlike entities as quasars, each apparently producing more than 100 times the energy radiated by the billions of stars in our galaxy, and pulsars, one of which was recently found to rotate 640 times in one second; the invisible "halos" of galaxies, containing as much as 90% of the galactic mass; threadlike structures more than a hundred light years long and a light year wide, cutting across the central region of our galaxy; "bubbles" in space, extending millions of light years, apparently devoid of stars; intergalactic and interstellar matter, including a variety of organic molecules; an apparent planet outside the solar system; probable "black holes" from

which light cannot escape; and the universe's "background radiation," hypothesized to be the remnants of the very first moments of creation.

2. Employing giant particle accelerators and electronic recording instruments so sensitive as to be able to detect the hypothesized decay of a single proton in a mass containing 10^{33} protons, physicists have discovered or experimentally produced large families of particles (in addition to the electrons, protons, and neutrons that are the constituents of ordinary matter) including anti-particles of the particles; investigated the internal structure of protons and neutrons; produced enormously massive particles tending to confirm theories unifying the gravitational and weak nuclear force; precisely measured the increase of particle mass and the slowing of time at velocities approaching that of light, as predicated by Relativity Theory; and, by inducing such phenomena as those of super-conductivity and those manifested in Josephson junctions, tended to confirm some of the wierder aspects of Quantum Theory.*

More important have been the advances in theory:

1. Relativity Theory, the last and greatest of classical theories in physics (classical in still assuming the continuum and determinism, which quantum theories have, in effect, discarded). Among its accomplishments: overcoming the anomalies (notably the deviation from calculated values in the perihelion of Mercury) that had plagued Newtonian gravitational theory; reconciling the invariance requirements of Newtonian theory with those of Maxwell's theory of electromagnetism; demonstrating the equivalence of gravitational and inertial fields, and of mass and energy; combining space and time and relating space–time to the matter and energy that structures it and determines its ge-

---

*Achievements almost as significant have been made in thermodynamics and other branches of physics, but they are peripheral to the concerns of this book.

ometry; predicting the generation of nuclear energy and the discovery of "black holes."

2. Quantum Theory in all of its ramifications, generally regarded as the most successful complex of scientific theories yet formulated, because it has survived every observational and experimental test so far devised and every argument leveled against it (including those of such formidable intellects as Einstein); the precision with which its predictions have been fulfilled; the enormous range of data it has been able to subsume and order; its permeation of other sciences, particularly chemistry, and branches of technology, notably electronics and related fields. Somewhat more specifically:

a. The formulation of Quantum Electrodynamics, which, among other contributions, explained why an atom's orbiting electrons do not fairly quickly collapse into the nucleus (as pre-quantum physics had required); elucidated atomic structures and, therewith, the Balmer series of spectral lines in atomic radiation; and worked out, with extraordinary precision, the interaction of electro-magnetic radiation with matter.

b. The quark model for hadrons (which includes mesons and hyperons as well as protons and neutrons) and the associated chromodynamics, which brought order to the "particle zoo," the seemingly endless proliferation of "elementary" particles, running into the hundreds, of the 1950s and 1960s.

c. The partial melding of Special Relativity and Quantum Theory in Quantum Electrodynamics and of General Relativity and Quantum Theory in theories that purport to explain the nature and presumed evolution of black holes.

d. Cosmological models (elaborated by computer) based on Relativity and Quantum theories, together with Thermodynamics, which purport to trace the history of the universe back some 15–20 billion years to the first 10^{-43} seconds after the hypothetical "Big Bang" of creation and forward either to a "heat death" in which all matter will have disintegrated, decayed, and dissipated, leaving only photons and neutrinos in an endlessly expanding universe; or, alternatively, to a reversal of the universe's expansion (perhaps some 60–80 billion years from now) to a re-coalescing "Big Crunch," in which all of the universe's matter, energy, space, and time may be contained in a point (a singularity) from which could emerge another Big Bang and a recreated universe.

   e. A number of competing Grand Unification Theories (GUTs) or "Theories of Everything" (TOEs), which have sought to show that the four basic forces or interactions (gravitational, electromagnetic, strong nuclear, and weak nuclear)* are all manifestations of symmetry-breaking in one primordial force. As indicated above, empirical findings have tended to confirm the theoretical association of the electromagnetic and the weak nuclear force in an electro-weak force. Empirical confirmation of the linking of the electroweak and the strong nuclear force is still lacking — it is currently being sought in evidence, not yet conclusively found, of proton decay. Some physicists (propounders of "supergravity" and related theories) expect that before very long even gravity (which has not yet been successfully quantized) will be brought within the scope of some ultimate GUT.

   Among those who share this expectation is the famous theorist of Black Holes and the Big Bang, Stephen Hawking. In his inaugural address (published by Cambridge University Press in 1980) as Lucasian Professor of Mathematics at Cambridge University, "Is the End of Theoretical Physics in Sight," he seriously advances "the possibility that the goal of theoretical physics [namely, an all-encompassing, unified, final theory of all microphysical and macrophysical

---

*In January 1986, evidence of a possible fifth basic force tentatively labeled "hypercharge" was presented and supported by reputable physicists.

phenomena] might be achieved in the not too distant future, say, by the end of the century."*

Yet, for all these triumphs (and the triumphalism they have evoked), there is widespread unease among physicists, as well as among philosophers of science and other scholars seriously concerned about understanding physical reality — the feeling that the conception of reality emerging from the recent progress of physics just doesn't make sense.

The Crisis of Intelligibility

Physics, probably more than any other science, has always been beset by crises of intelligibility, by paradoxes contrary to received opinion, antinomies contradictory to "established" natural law, mysteries impenetrable to common sense, seeming absurdities.

Thus, as Ptolemy and his followers superimposed epicycles upon epicycles to reconcile their models with the perceived movement of the planets, the feeling grew that the Creator could not have worked in such circuitous ways. The alternative Copernican model was seen as contrary not only to Holy Scripture but to the common and incontrovertible perception that "the sun riseth and goeth down." Galileo's theory of gravitation violated the universally accepted Aristotelian notion that bodies of different weight must fall to earth at velocities proportional to their weights. The Newtonian theory disturbed natural philosophers, most of whom found it hard to accept the "action at a distance" that

---

*This brings to mind E. Haeckel, who in his widely hailed and influential work *Die Welträtsell* (1889), translated by J. McCabe as *The Riddle of the Universe at the Close of the Nineteenth Century* (New York and London: 1901), maintained that *all* the fundamental problems of *all* the sciences had already been solved or were on the verge of solution — this, in the face of the critical difficulties that were, at the beginning of the twentieth century, to lead to the Relativity and Quantum revolutions.

the theory postulated and that Newton himself (though confessedly troubled by it) never tried to explain ("*hypothesi non fingo*"). The demonstration in thermodynamics that "time has an arrow" was profoundly offensive to classical physicists, who firmly held to the view that physical processes and the equations that describe them must be time-symmetrical. Relativity Theory, with its curved space, dilated time, increasing mass at velocities approaching that of light; its rejection of the commonsense conviction that the perceived velocity of light had to be affected by movement toward or away from its source; and all the other strange ideas it harbored or generated, met fierce resistance from conventional physicists and philosophers. So, too, did the early quantum theories of Planck and Einstein, entailing, as they did, the rejection of the continuum.

In time, these difficulties were resolved. Aristotelian and Ptolemaic concepts were simply discarded (except by religious fundamentalists and still-continuing believers in a "flat earth" and other such ideas). Newton's *Principia* corrected and systematized the work of Copernicus, Kepler, and Galileo and became the new orthodoxy. The bothersome "action at a distance" was tacitly accepted and then explained, after a fashion, by resort to field concepts adapted from Maxwell's theory of electromagnetism. Boltzmann's statistical mechanics brought thermodynamics sufficiently into line with classical concepts to satisfy most physicists and philosophers of science. Relativity Theory benefited from a massive effort (in which Einstein himself participated) to assimilate it to common sense. Similarly, with early quantum theory, ideas that had seemed strange, therefore repugnant, gradually became acceptable through growing familiarity. Scientists and philosophers who couldn't accept the new view of reality died off to make way for those not committed to old preconceptions.

The present crisis of intelligibility — which began in the mid-1920s, with the formulation and elaboration of quantum mechanics by Bohr, Schrödinger, Heisenberg, Born, de

Broglie, Pauli, Dirac, and others — differs fundamentally from all the others. The problem is not merely that of the unfamiliarity, strangeness, or heretical character of the theory, nor its failure to explicate the "how" of phenomena — that was also true, as noted, of Newtonian theory — but the theory appears to be illogical and self-contradictory, to be involved in and to lead to absurdities.

There are first the queer ideas that keep cropping up — some later to be muted or abandoned, others persisting through various transformations: negative energy and negative mass; time reversal (as in the famous Feynman diagrams in which positrons are conceived as electrons traveling backwards in time); the many universes, perhaps infinitely many, supposed to be created at each point-event (all the possibilities not realized in this universe, as each space–time point, being realized in other universes, with which communication is impossible); that the smaller can contain, and accordingly be large than, the larger (thus a proton is supposed to contain within itself "messenger" particles 10^{14} times more massive than itself; an electron posited as having an infinite self-energy and therefore presumably having infinite mass — a contradiction to empirical fact disposed of by the artificial mathematical procedure of "renormalization," in which the infinite terms in an equation are simply arbitrarily removed*; that leptons, supposed to be dimensionless points, can have masses varying from that equivalent to 1.8 billion electron volts to less than 1000 electron volts; that the vacuum, by definition empty, devoid of matter, is actually a plenum seething with "virtual" particles (particle–anti-particle pairs) that are supposed vastly to outnumber actual particles; that everything in the universe, not only matter–energy but space–time, emerged out of the sin-

---

*This is but one example, though the most notorious and troublesome, of the practice of arbitrarily introducing and imposing variables, constants, parameters, or sleights of hand into theories to make them fit observed phenomena — the "force-fitting" of theory.

gularity of the Big Bang, a dimensionless point, a nothing; and that at the Big Crunch everything will again become nothing, "not only matter, but the space and time that envelops that matter,"* in plain contradiction to the conservation laws that virtually all physicists dogmatically accept.

Then there are the enormous mathematical and logical difficulties in current theory, not only the problems with renormalization just mentioned, but that the probability theory and the statistics at the heart of Quantum Theory do not conform to classical probability calculus with its Kolmogoroff axioms, the Boolean lattice of events, etc., which all physicists are still taught; and that Quantum Theory, though built up by means of classical logic, is not consistent with it.

Of course, quantum theorists have a solid justification for such departures from classical logic, mathematics, and statistics: the facts observed dictate the departures. And they have an explanation, *of a sort,* for the paradoxes that appear in the theory, that provided, for example, by Heisenberg's Uncertainty Principle, which permits the quantities of mass required by theory to come into being for infinitesimally short time intervals. (Since the uncertainty of a particle's momentum — the product of its mass and its velocity — approaches infinity as its position becomes progressively more definite, momentum, and accordingly mass, can be assumed to be as great as required during a period defined by Planck's constant.)

However, such explanations do not suffice for the most crucial set of paradoxes and enigmas that today characterize quantum theory: those deriving from demonstrations that what happens at a given location is affected, *at velocities exceeding those of light, and perhaps instantaneously* — in flat contradiction to a central tenet of Relativity Theory — by what happens at some distant location; that a present act of

*J. A. Wheeler, in *Some Strangeness in the Proportion,* a Centennial Symposium to Celebrate the Achievements of Albert Einstein, ed. by H. Wolf (Reading, PA: Addison-Wesley, 1980).

observation will ineluctably determine what happened in the past, even billions of years ago; and that in general *objective reality* depends on *subjective observation*. In Alain Aspect's 1982 experiment (and in subsequent tests) showing that "Bell's Inequality" is violated, the polarization at a calcite detector of one of two paired photons moving in opposite directions is influenced by the polarization of the other photon some distance away. Moreover, distance doesn't appear to matter. The distance might be a few meters and the time lapse a hundredth of a microsecond; but the distance might just as well be billions of light years and the time lapse accordingly billions of years. Thus, present observations, determined by the last-minute setting of a calcite or other detector, have decisive consequences for a photon that may have been emitted from a far-distant star billions of years ago. This is apparently shown by observations made of two astronomical objects, known as 0957 + 561 A,B, once thought to be two separate quasars, but now widely considered to be two images of the same quasar (the bifurcation of images being caused by the lensing effect of an intervening galaxy).

As J. A. Wheeler comments in a paper read at a joint meeting of the American Philosophical Society and the Royal Society, on June 5, 1980: "An astronomer arbitrarily decides," about an incoming photon from the distant quasar, whether to observe by "which route" or to observe interference between "both routes". . . . When it triggers a counter, we discover "by which route" it came with the one arrangement; or, by the other, what the relative phase is of the waves associated with the passage of the photon from source to receptor "by both routes" — perhaps 50,000 light years apart as they pass the lensing galaxy G-1. But the photon has already *passed* that galaxy billions of years before we made our decision. This is the sense in which, in a loose way of speaking, we decide what the photon *shall have done* after it has *already* done it (Wheeler's emphasis). In short, according to quantum theory, "the past has no existence except as it is recorded in the present" [J. A. Wheeler: "The 'past' and the 'delayed choice' double-slit experiment," in *Mathematical Foundations of Quantum Theory*, edited by A. R. Marlow (New York: Academic Press, 1978)].

If, as N. Bohr, the founding father of the new quantum theory, argued, the act of observation determines reality, not only does the present determine the past, but future observations will also determine the present reality, occasioning "bizarre paradoxes of causality in which observers in some frames of reference find that one event is 'caused' by another that has not yet happened" (B. d'Espagnat, in *Scientific American*, November 1979).

That is why Wheeler speaks of a "participatory universe," a phrase he has made famous. However, in his contribution (*op. cit.*) to *Some Strangeness in the Proportion*, he admits the irresolvable paradox involved in such observer participation: "'The universe' exists 'out there' independent of acts of registration, but the universe does not exist 'out there' independent of acts of registration."

Attempts at Resolution

As might be expected, many physicists and other theorists have tried to resolve such paradoxes, to dispel such mysteries. Some, including D. Finkelstein, B. Van Frassen, R. J. Greechie, and R. F. Cudder, have proposed a variety of non-Boolean "quantum logics," intended fundamentally to alter not only *what* we think of physical reality, but *how* we think and speak of it. However, these efforts run afoul of Bohr's stricture that all experiments, all observations, all theories must ultimately be referred back to the ordinary common-experience, commonsense way of thinking and speaking: "All experiments must ultimately be expressed in terms of classical concepts" (quoted in A. Peterson, *Quantum Physics and the Philosophical Tradition*, Cambridge, MA: MIT

Press, 1968). Otherwise, the ideas of quantum theory will be further shrouded in mystery and obscurity. For this and other reasons, quantum logicians have had scant success and still less acceptance. None appears to offer a genuine resolution of the conceptual difficulties that now plague Quantum Theory. Their efforts to restore intelligibility are themselves unintelligible.

More serious have been the efforts of groups of physicists, led respectively by D. Bohm and J. P. Vigier, who have offered an assortment of "hidden variable" theories. While the quantum logicians have been concerned with the way we think and speak about physical reality, the hidden variable theorists have been concerned with the nature of the underlying reality itself. They have variously argued that, underlying observed particle states, and determining them, is an unobserved and probably unobservable field or medium from which the observed states emerge. Nobel laureate L. de Broglie, in his guarded introduction to *The Vigier Theory of Elementary Particles* (Amsterdam and New York: Elsevier, 1963) presents one conception of the hypothetical field:

> The subquantic medium would be made up of an entirely chaotic wave field containing innumerable small regions of high concentration of the field (singular regions) and storing up a formidable quantity of hidden energy. . . . This medium might be conceived of as a sort of gas, undoubtedly made up of leptons and perhaps of neutrinos which would move isotropically in every direction with all possible energies . . . But why would this subquantic medium containing such a formidable reserve of energy. . . . totally elude observation. Undoubtedly because of its totally chaotic character which has as a result an average zero value for all the quantities involved. Nevertheless, in this chaotic medium it would be possible for the particles making up the medium to form various combinations . . . and in this way more complex particles would form in the very heart of the medium. Only a few of the innumerable centers of concentration of the field

would be able to "emerge" from the subquantic medium, and organize around themselves, by giving it a rhythm, an extended coherent wave that would be propagated as a slight perturbation on the "surface" of the sub-quantic medium. Escaping in this way from the characteristic chaos of the medium, these particles would become indirectly observable by us on the macrophysical level. . . .

While this picture prefigures current theories of the dense clouds of "virtual particles" that are said to envelop every actual particle, and while J. S. Bell and others have shown that the long-accepted "proof" by J. von Neumann (in his *Mathematical Foundations of Quantum Mechanics*; Princeton University Press, 1955) that quantum theory does not permit hidden variable interpretations is based on questionable assumptions, the vast majority of physicists have rejected, or just ignored, hidden variable theories, including the increasingly sophisticated ones that the theorists keep offering.

The reasons for the fading of hidden variable theories are compelling.

One difficulty is that all the leading versions of the theory require causative influences transmitted faster-than-light, which is expressly excluded by Relativity Theory.

A related difficulty is that the theories offered are as strange, as alien to common sense, as the standard theory they are intended to replace. And they are becoming ever more complicated, forbiddingly so.

Perhaps the most important problem arises from the consideration that the variables postulated are hidden. Since we can know nothing directly about them, a great many theories about their nature and behavior have been propounded and a great many more are conceivable. By suitably adjusting the properties of the postulated field, medium, pilot wave, or whatever, "one can always invent hidden variable models that can reproduce the statistical results" (C. A. Hooker, in *Paradigms and Paradoxes*, edited by R. G. Colodny, University of Pittsburgh Press, 1972). D.

Bohm admits as much (*Causality and Chance in Modern Physics*, Princeton, NJ: Van Nostrand, 1953). The only way to choose among the theories — and, what is more to the point, giving the one chosen scientific respectability — is to predict empirical findings not in conformity with standard quantum theory in its Copenhagen interpretation, and having such predictions confirmed by experiment and observation. The trouble is that no such findings have so far been predicted, let alone confirmed. There is accordingly no real difference between hidden variable theories and orthodox quantum theory. They are empirically equivalent. They differ only in that the hidden variable theories have a good deal of what has been termed "metaphysical baggage" of an essentially arbitrary, speculative character, lacking in elegance, and not in itself intelligible.

In conclusion, both the quantum logic and the hidden variable theories are fringe phenomena, rejected or ignored by mainline physicists, and incapable of resolving the crisis of intelligibility in present-day physics.

Despair of Intelligibility

The reaction of most physicists to the paradoxes and enigmas of quantum theory, and to the failure to resolve them, has been to despair of ever being able to make the theory intelligible.

Many years ago, before the mystery had deepened, Nobel laureate P. W. Bridgman poignantly voiced the sentiment: "Our conviction that nature is understandable and subject to law arises from the narrowness of our horizons. . . . If we sufficiently extend our range, we shall find that nature is intrinsically and in its elements neither understandable nor subject to law" (*Reflections of a Physicist*, New York: Philosophical Library, 1950).

Another Nobel laureate, R. Feynman, gave this advice to his students in his famous lectures of the 1960s: "I think it is safe to say that no one understands quantum mechanics. Do not keep saying to yourself, if you can possibly avoid it, 'But how can it be like that?' because you will go 'down the drain' into a blind alley from which nobody has yet escaped. Nobody knows how it can be like that."(R. P. Feynman, R. B. Leighton, M. Sands, *The Feynman Lectures on Physics*, 3 volumes, Reading, MA: Addison-Wesley, 1965.)

More recently, J. A. Wheeler has led a swelling chorus in the same basic melody. Typical of his utterances:

> There is no law of physics which does not lend itself to most economical derivation from a symmetry principle. However, a symmetry principle hides from view any sight of the deeper structure that underpins that law and therefore also prevents any immediate sight of how in each case that mutability comes about. Moreover, no search has ever disclosed any ultimate underpinning, either of physics or of mathematics, that shows the slightest prospect of providing the rationale for the many-storied tower of physical law.

("Genesis and Observership," in *Foundation Problems in the Special Sciences*, edited by Butts and Hintikka, Dordrecht, Holland, and Boston, MA: D. Reidel, 1977.)

To such despair of understanding, there have been three kinds of responses. The first is the traditional rejection of the very effort to explain, to comprehend, to make intelligible, an attitude that has its roots in religion and in the point-of-view of practical people who are not interested in the "why" of the world but only in the "how." The classic expression of this attitude among scientists is, of course, the well-known passage of Newton's *Principia*:

> I have not been able to discover the cause of those properties of gravity from phenomena and I frame no hypotheses [*hypothesi non fingo*]; for whatever is not deduced from the phenomena is to be called an hypothesis, and hypotheses, whether physical or metaphysical, whether of occult qualities or mechanical, have no place in experimental philosophy. . . . And to us it is enough that gravity really exists and

acts according to the laws which we have explained, and
abundantly serves to account for all the motions of the ce-
lestial bodies, and of our sea.

Bohr, the primary formulator of the still-dominant Copenhagen interpretation of quantum theory, expressed
himself similarly (as quoted by A. Petersen, *op. cit.*): "There
is no quantum world. There is only an abstract quantum-
physical description. It is wrong to think that the task of
physics is to find out how nature is. Physics concerns only
what we can say about nature."

Nobel laureate E. Schrödinger, who along with M.
Planck, L. de Broglie, and other creators of quantum theory
had shared Einstein's passionate conviction "I still believe in
the possibility of a model of reality — that is, of a theory
which represents things themselves and not merely the
probability of their occurrence" ("Autobiographical Notes,"
in *Albert Einstein, Scientist–Philosopher*, edited by P. A.
Schilpp, Evanston, IL: Library of Living Philosophers,
1949), came eventually to agree with Bohr: "Reality resists
imitation through a model. So one lets go of naive realism
and leans directly on the indubitable proposition that *actually* (for the physicists) after all is said and done, there is
only observation, measurement. Then all our physical
thinking thenceforth has as its sole basis and as sole object
the results of measurement which can in principle be carried out, for we must now explicitly *not* relate our thinking
any longer to any other kind of reality or to a model"
(Schrödinger's emphasis; from a translation of one of his
1935 papers originally published in *Die Naturwissenschaften*,
in *Quantum Theory and Measurement*, edited by J. A. Wheeler
and W. H. Zurek, Princeton University Press, 1983).

So, too, P. Wigner, in his Nobel Prize lecture of Decem-
ber 1963 (reprinted in *Science*, September 4, 1964): "It is
often said that the objective of physics is the explanation of
nature. . . . [i.e.] the establishment of a few simple principles which describe the properties of what is to be explained. . . . In this sense, physics does not endeavor to ex-

plain nature. In fact, the great success of physics is due to a restriction of its objectives; it only endeavors to explain the regularities in the behavior of objects."

In our own day, Nobel laureate I. Prigonene starkly expresses the still prevalent view: "As for the scientist, his concern with nature would be reduced to taking it as a set of manageable and measurable objects; he would thus be able to take possession of nature, to dominate and control it but not understand it. Thus the intelligibility of nature would be beyond the grasp of science" (*Order Out of Chaos, Man's New Dialogue with Nature,* New York and London: Bantam Books, 1984).

Today, the overwhelming majority of physicists are unconcerned with theories that purport to explain phenomena, but appear to offer no gain in uncovering or producing new phenomena, refining their measurement, or testing hypotheses about how the phenomena relate to each other. Whether complacently or despairingly, they acknowledge that quantum theory is not a means of understanding nature but only a set of rules and procedures for measuring, calculating, and relating phenomena that somehow emerge from an unknowable nature to become objects of observation, measurement, and other aspects of the scientific discipline.

Traditionally, philosophy of science and such other branches of philosophy as ontology, cosmology, and epistemology have sought that understanding of nature of which scientists despair or to which they are indifferent. But philosophers have also, for the most part, in effect decided that understanding is beyond their reach also. Ever since L. Wittgenstein repudiated his effort in *Tractatus Logico-Philosophicus* (London: Routledge & Kegan Paul, 1922) to construct a comprehensive and meaningful "picture of the world" from "atomic facts" and turned his attention instead in *Philosophical Investigations* (Oxford: Basil Blackwell, 1953) and other works to "word games," for which he did not claim much importance — and indeed they have little — the great majority of philosophers in the Anglo-Saxon tradition have followed his course into thickets of linguistic analysis, in which the practitioners can hardly find each other, let alone find anything that can help make sense of the world to the rest of us.*

Even on the European continent, where the old metaphysical pretensions of traditional philosophy — the claim of being able to construct ontologies and cosmologies, largely, perhaps entirely, unrelated to technical science, that might enable us to understand reality — the tradition has worn thin. The Bergsonian, phenomenonological, and existentialist philosophies that dominated European thought during the first half of the twentieth century were concerned with our apprehension of the world rather than with the nature of the world apprehended. And these philosophies have given way to structuralist and deconstructionist schools, increasingly technical and artificial, increasingly remote from common sense. In deconstructionism, now strongly in vogue, the very effort to achieve intelligibility is derided; and the classic endeavors of philosophy have been reduced to intellectual (and often silly or outrageous) games, different, to be sure, from the word games of the Anglo-Saxon tradition, but no more related to reality and the effort to make it intelligible, no less otiose and trivial. As to "objective reality," the dominant attitude is that of Wittgenstein's famous dictum in the *Tractatus*: "Whereof one cannot speak, thereof one must keep silent."

The second main response to the crisis of intelligibility

---

*Mention must, however, be made of certain exceptions, notably Brand Blanshard, one of the last survivors of the Hegelian idealism that dominated Anglo-American philosophy in the last quarter of the nineteenth century and the early years of the twentieth. Blanshard still insists on the intelligibility of reality, on its complete accessibility to reason. Everything, he argues, can be explained. Everything can, at least in principle, be shown, by thinking alone, to have had to be what it is. Alas, the thinking of this venerable philosopher has had few echoes among his younger colleagues.

in physics has been a quest for new modes of understanding in mysticism and a variety of oriental philosophies, of which *The Tao of Physics* by F. Capra, *The Wu-Li Masters* by G. Zukov, and *Star Wave* by F. A. Wolf are representative and which such reputable physicists as D. Bohm and D. Finkelstein have endorsed. But since this course means the abandonment of the quest for intelligibility in terms acceptable to western secular thought, this tendency is, in effect, a reaffirmation of despair of intelligibility.

There is still a third kind of response to the crisis of intelligibility. Recognizing that observation, measurement, prediction, and control, without comprehension of what is being observed, measured, predicted, and controlled, is essentially only technology and is indeed tantamount to magic; and that its practitioners become *fachidioten*, "jobbers and fragments of humanity," of which Nietzsche prophetically wrote almost a century ago;* recognizing, too, that denial of an objective reality, accessible to thought but independent of it, is favorable to totalitarianism, as G. Orwell in his *1984* and K. Popper in his *The Open Society and Its Enemies* warned, a small but growing band of eminent physicists have called for a resumption of the struggle to which Einstein devoted the last decades of his life, the rejection of the subjectivism inherent in the Copenhagen interpretation of quantum theory, the search for an objective reality independent of our acts of observation, the endeavor to make that reality intelligible. And this, it is more and more widely recognized, requires not merely new equations, but a new

---

"For seventeen years I have never tired of calling attention to the despiritualizing influence of the current science–industry. The hard helotism to which the tremendous range of the sciences condemns every scholar today is a main reason why those with a fuller, richer, profounder disposition no longer find congenial education nor congenial educators. There is nothing of which our culture suffers more than of the superabundance of pretentious jobbers and fragments of humanity; our universities are, against their will, the real hothouses for this kind of withering of the instincts and of the spirit" ("Twilight of the idols,"* in *The Portable Nietzsche*, edited by W. Kaufman, New York: Viking, 1954).

paradigm, a "new simple idea" (J. A. Wheeler's words, in a television broadcast on November 20, 1985).

Thus, Abdus Salam, a Nobel laureate for his contribution to GUT: "All science — physics in particular — is concerned with discovering *why* things happen as they do. The 'why's' so adduced must clearly be 'deeper,' more universal, more axiomatic, less susceptible to direct experimental testing, than the immediate phenomena we seek to explain" (in *Scientific Explanation*, edited by A. F. Heath, Oxford: Clarendon Press, 1978).

J. A. Wheeler, in his "Genesis and observership" (Butts and Hintikka, *op. cit.*): "It is not the way of science to sit inactive in the face of mystery." And from his preface to *Quantum Theory and Measurement* (*op. cit.*): "And from what deeper principle arises the necessity of the quantum in a construction of the world"; and from a comment on p. 210 of that compendium of papers: "Today we demand of physics some understanding of existence itself."

So, too, and more fully, J. S. Bell, the formulator of the "inequality," the violation of which is a leading influence in the generation of the paradoxes and mysteries noted above:

> The awkward fact remains that [quantum] theory is only *approximately* unambiguous and *approximately* self-consistent. . . . It is interesting to speculate on the possibility that a future theory will not be *intrinsically* ambiguous and approximate. Such a theory would not be fundamentally about "measurements," for that would again imply incompleteness of the system and unanalyzed intervention from the outside. Rather it should again be possible to say of a system not that such and such may be *observed* to be so but that such and such *be* so. The theory would not be about *observ* able but about *be* able.

(His emphasis; from his contribution to *The Physicist's Conception of Nature*, edited by J. Mehra, Reddel Publishing Co., 1973.)

* * *

Is such a theory possible? Is it obtainable, constructable? Can the paradoxes and mysteries that today obscure physicists' view of reality be resolved and dispelled? Can physics be made intelligible? Can the world described by physics be made intelligible?

To answer these questions, we must first determine the conditions necessary for, or at least conducive to, intelligibility.

Conditions for Intelligibility

Three views of these conditions are currently prevalent: that tacitly held by the general public; that dominant among physicists of essentially classical persuasion and among educated and thoughtful laymen; and that widely entertained by philosophers of science and those quantum physicists who concern themselves with the problem.

1. For most people who are neither scientists nor philosophers, the main condition for intelligibility is familiarity. What is imbibed, as it were, "with mother's milk" is not questioned, is taken for granted, implicitly understood.

Yet familiarity is neither sufficient nor necessary for intelligibility. For example, the fact that water turns to ice at 0°C and to vapor at 100°C, something so familiar as rarely to excite wonder, let alone bewilderment, does not in itself make that fact intelligible. That water molecules, as they lose velocity with falling temperatures, would approach immobility, and move more and more wildly about as they gain velocity with rising temperatures, may be intuitively understandable. But just why the 0°C and 100°C temperatures for the phase transitions? Why are the transitions relatively abrupt, rather than gradual? Physical chemistry has, to be sure, partially answered such questions, but the answers are embedded in complex theories that are not at all familiar to most people and that they would not be able to understand without intensive study. That a teacup doesn't fall to the floor or ground requires, for most people, no other explanation than that it is supported by a shelf, a table, a hand, or whatever; but they would have difficulty understanding, let alone explaining, how an object almost entirely composed of empty space can support another. A scientific explanation would require resort to unfamiliar notions, often of daunting complexity and perplexity, about the nature of electrical and magnetic forces, Pauli's exclusion principle (which is basic to atomic and molecular structures), the special characteristics of amorphous and crystalline solids, the physiology of muscular action, etc.

Brute facts, however familiar, are never in themselves intelligible. They have to be made intelligible. Only theories or other conceptual frameworks in which they are subsumed can make them so — but only if these theories or frameworks are themselves intelligible.

By the same token, the familiarity of a theory and the sense of intuitive clarity and verisimilitude it tends to induce do not, in themselves, make the theory intelligible. It may be true, as Einstein held, that "most of the essential ideas of science are essentially simple and may, as a rule, be expressed in language comprehensible to everyone"; but what, if "the ideas . . . made comprehensible to everyone" are, when critically examined, enigmatic? Bohr's famous dictum that "all experiences must ultimately be expressed in terms of classical concepts" and accordingly in terms consonant with common experience and common reason, may satisfy Schrödinger's requirement: "If you cannot — in the long run — tell everyone what you have been doing, your doing it has been worthless." But what if while what physicists have been doing with their particle accelerators, their radio telescopes, and all their other apparatus can be clearly described in terms quite intelligible to the average reader of *Popular Mechanics,* say, the results obtained are still paradoxical, puzzling, mysterious, as indeed some of them are?

Popularizers of current physical theory like C. Sagan, I. Asimov, H. Pagels, N. Calder, P. Davies, among a host of other adepts at lucid exposition, have made the procedures, concepts, and theories of contemporary physics familiar to hundreds of thousands, maybe millions, of readers and television viewers — with the result that such concepts as "curved space," time as a fourth dimension, $E = mc^2$, the Uncertainty Principle, Black Holes, the Big Bang, GUTs, have become clichés of the mass media and of cocktail party chit-chat. However, at bottom, the popularizers have merely exposed and celebrated the mysteries that pervade physics, rather than explained them. They have translated mathematical abstractions that, by their nature, preclude formulation in ordinary language concepts acceptable to common sense, into enticing but misleading metaphors.

Just as familiarity doesn't suffice for intelligibility, so, too, it is not necessary for it. It could well be that ideas that might make physical reality understandable to penetrating and rigorous thinkers might still leave the ordinary person unenlightened and perplexed. Most people require images, diagrams, mechanical models of some kind — or verbal descriptions that evoke such percepts — to make phenomena comprehensible. That is why Lord Kelvin insisted that physical theory be representable in terms of rigid rods, balls, levers, and other such elements of mechanical models. But such models have been rejected in this century as not truly representative of a deeper reality, whether micro- or macro-physical. A space–time manifold of more than four dimensions is not visualizable, yet is quite intelligible, as a mathematical abstraction, and might well be an aspect of, or even be central to, an intelligible theory of physical reality — provided we are able to explain why physical reality is characterized by ten or eleven dimensions, as some theorists now conjecture, rather than by one hundred or by two; and provided also that attribution of ten or eleven dimensions to reality resolves the paradoxes, dispels the mysteries,

of contemporary physics. On the other hand, a miracle —
e.g., the splitting of the Red Sea, or Shadrach, Meshach,
and Abednego remaining quite comfortable in a fiery fur-
nace, or Jesus walking on water — are all readily visualizable (and indeed they have been entertainingly visualized
by Cecil B. deMille and other film directors); yet they cannot be regarded as intelligible by scientific or other rational
criteria.

2. For people still under the spell of classical physics,
and the mechanical world-view associated with it, strict, linear determinism is necessary for intelligibility. Every effect
(e.g., the movement of an electron, the fission of an atomic
nucleus) must have a determinate and fully and precisely
specifiable cause; every cause must have determinate and,
in principle, specifiable effects. In principle, a coin falls
heads instead of tails for definite and sufficient reason,
even though we cannot, in practice, determine the reasons.
Because this condition is not satisfied in quantum theory —
indeed, classic determinism is categorically rejected by the
theory in all its variants and progeny — that theory (even
in its earlier, and less problematical, formulations) is just not
intelligible to classic-minded people. It needs "completion,"
as Einstein insisted, to make it intelligible.

However, it is quite understandable that the probability
of a coin — at least, an abstract coin, perfectly fair and unaffected by such influences as biased spin, an uneven table
surface, a prevailing wind, or whatever — falling heads
should be 1/2; and that the probability of a die (fairly fab-
ricated and fairly cast) falling with the numeral 2 upwards
should be 1/6. Nor is it unintelligible that a sequence of
events should be random (though there are still unresolved
difficulties about determining whether or not a sequence is
truly random, that it is *not* actually governed by some rule,
however complex or subtle); and that the occurrence of any
event in the sequence should therefore be unpredictable.

The overwhelming majority of physicists (and of other scientists and philosophers of science) today accept probabilistic–statistical determinism as intelligible, even if not as esthetically satisfying (in part because of its simplicity) as classic determinism. In that sense, quantum mechanics *is* deterministic. The probabilistic wave function of a system, its equations of state, at any given moment, *determines* its wave function at later moments. Quantum indeterminacy restricts what may be *measured* in any given state, but *not* the way the state evolves, which remains deterministic.

What is *un*intelligible is not that "God plays dice," but that he appears to be doing so in a "black box." Einstein eventually came to accept that God does indeed play dice, but he objected (as quoted in Wheeler & Zukov, *op. cit.*) "That the Lord should play with dice, all right; but that he should gamble according to definite rules, that is beyond me." Actually, his difficulty — a difficulty still not overcome — was that these rules seemed arbitrary. True, the psi function, which governs the statistical distribution of sub-atomic phenomena, derives from the fundamental quantum of action; but not only do we not now understand why there are discrete quanta (instead of a continuum) and why the quantum has the value it has, but there is no agreement about the physical meaning of the psi-function. A cloud of moving particles behaves like a wave merely because the statistical probability of finding a particle at any given position is representable by a wave equation. Why just such an equation rather than some other, has not been explained. Mysterious also are the reasons why certain particles (electrons, protons, neutrons, quarks, etc.), those subject to Pauli's Exclusion Principle, are governed by Fermi–Dirac statistics, while certain other kinds of particles, not subject to Pauli's Principle (photons, pions, K mesons, etc.) are governed instead by Bose–Einstein statistics.

The fundamental difference between an intelligible and an unintelligible statistics is that in the former the mechanism underlying the statistical distribution is manifest; in the latter, it is not. It is the difference between understanding the statistical distribution of the symbol combinations appearing on the window of a slot machine from knowing just how the machine has been constructed, how preset, how acted upon, how it might be affected by "tilting" or other influence, and "understanding" that distribution solely from analysis of past distributions, without knowing anything about the inner mechanism and the operating conditions underlying and determining the distribution. In the latter case, the probability function derives only from certain end appearances, thus furnishing no basis for accounting for the particular character of the probability function, nor for predicting departures from that function in future distributions. Since we don't know *why* the psi function, it is unintelligible. Quantum theory, in general, is unintelligible, not because its determinism is probabilistic rather than classical, but because its statistics is not intelligibly grounded.

Just as classical determinism is not necessary for intelligibility, so too is it not sufficient for it. Though classical mechanics is, in principle, thoroughly deterministic — so much so that, as in LaPlace's famous boast, if he knew with absolute precision the position and momentum of every particle in the universe, he could precisely calculate each particle's course forward to the end of time and backward to the Creation, and therewith every moment of the universe's history — it is not therefore intelligible. The action at a distance of the gravitational force — the only one that classical mechanics dealt with — was a mystery and so were many other aspects of classical deterministic mechanics. The radiance of the classical ideal was so great that it blinded the eye to the crucial difficulties entailed by the theory.

3. The condition for intelligibility regarded as both nec-. essary and sufficient by most contemporary philosophers of

science, and by those physicists and other scientists who give
any thought to the issue at all, is that phenomena be deriv-
able from, be covered by, be subsumed under some universal law. [This condition was first authoritatively formulated
by C. G. Hempel and G. Oppenheim in "Studies in the
Logic of Explanation" (*Philosophy of Science* **15**, 1948) and
has since then been variously elaborated and modulated by
Hempel and many other writers.] Thus, the movements of
the tides and the planets are explained by derivation from
gravitational laws (whether Newtonian or relativistic); the
orbits of electrons around an atomic nucleus by derivation
from Bohr's model of the atom and, more particularly,
from Pauli's Exclusion Principle and the laws of quantum
electrodynamics as formulated by Dirac, Feynman, and others; the fact that the momentum of two particles having a
common source but traveling in opposite directions, without possibility of influencing or communicating with each
other, is correlated (however distant they may be from each
other at time of measurement) is explained by the law of
the conservation of linear momentum, together with the
Uncertainty Principle and related laws and principles of
Quantum Theory; and so for all other phenomena supposed to be governed by law.

However, derivation from a law, however universal,
however well established, does not make a phenomenon intelligible if the law is not itself intelligible. Regularity, order,
pattern, lawfulness *are* fundamentally characteristic of nature. That is why mathematics, which is the study of types
of order, can be successfully applied to nature. That is why
laws — which are essentially sets of rules that permit predicting the outcome of experiment and observation — apply. But then the questions arise: Why a particular law
rather than some other? And, more to the point, what is the
causal mechanism, what are the actual physical events, underlying the law? What happens between two noncommuni-
cating particles that "compels" the correlation of their mo-

menta? What physical events between a quasar's emission of a photon five billion years ago and an observer here on earth today permits the present act of observation to determine what happened five billion years ago and billions of light years away? Quantum theory has succeeded amazingly in ordering an enormous range of phenomena, in formulating laws covering the behavior of electromagnetic radiation, sub-atomic particles, atoms, molecules, crystals; making possible the calculation, prediction, and control of such behavior. But it has not made the behavior intelligible, because (as almost all quantum physicists would today admit) the laws, as formulated, are not themselves intelligible. It follows that derivation from law is not a sufficient condition for intelligibility. For a law to be made intelligible, there has to be understanding of the underlying mechanism, whether deterministic or probabilistic, whether visualizable or not. Unless we understand the causal connections manifested in law, the law is not intelligible.

As B. d'Espinat commented (in *Scientific American*, November, 1979):

> Whenever a consistent correlation between events is said to be understood or to have nothing mysterious about it, the explanation offered always cites some link of causality. Either one event causes the other or both events have a common cause. Until such a link has been discovered, the mind cannot rest satisfied. Moreover, it cannot do so even if empirical rules for predicting future correlations are already known. A correlation between the tides and the motion of the moon was observed in antiquity, and rules were formulated for predicting future tides on the basis of past experience. The tides could not be said to be understood, however, until Newton introduced his theory of universal gravitation.

Unfortunately, as noted, that theory required "action at a distance," which Newton admitted he could not understand or explain. And relativistic gravitation, which has replaced Newtonian, itself still awaits unmistakable evidence of gravity waves and of gravitons as carriers of the gravitational force for the fulfillment that physicists seek of its explanatory potential. (Whether the needed evidence can be found and if found will possess the sought-for explanatory power will be considered in the final chapter.)

Nor is derivation from a law a necessary condition for intelligibility. The randomness of phenomena becomes intelligible on the assumption that there is *no* law governing the phenomena. Thus, the sequence of plays at roulette, the sequence of prices on the stock exchange or other "perfect market," the utterances of a madman, become intelligible if we assume that they are governed by some effective randomizing device — a perfectly balanced and fairly spun roulette wheel; the complete, precise, and instantaneous discounting of stock market prices in the light of all information pertinent to the valuation of corporate shares; the derangement or disruption of the nervous circuitry governing orderly speech. There are, to be sure, boundaries to the range of possibilities — the number 58 or the color green cannot appear on the usual roulette wheel; the price of an IBM share cannot reach $100,000 in any one day; the madman cannot speak like a whale — but there is no law within the boundaries. Yet the phenomena are intelligible.

The inescapable conclusion, I submit, is that none of the conditions for intelligibility — familiarity, classical determinism, derivation from law — commonly proposed (or tacitly entertained) are necessary or sufficient for intelligibility.

* * *

What then *are* the conditions for intelligibility?

The basic condition is that it be possible to represent, model, map, what is to be explained (the explicandum) on an explicans that is itself intelligible, and that is, in significant respects, isomorphic, at least homomorphic, with the explicandum.*

What has to be explained are the phenomena appearing to physicists in their observations and experiments. Quantum Theory, it is generally admitted, has failed in this task; it has no clear, acceptable explanation of the most crucial of the phenomena it covers.

Ideally, the appearances would be mapped onto objective reality, the reality presumably underlying appearances. Unfortunately, objective reality cannot be directly known. Kant's *ding an sich* is forever out of reach, despite the claims of certain mystics. Recognizing this, many physicists and philosophers of science have been led to the revival of the view (of which Bishop Berkeley gave the most elegant formulation in the eighteenth century) that the act of observation determines the universe. As J. A. Wheeler muses (in *Genesis and Observership, op. cit.*):

---

*It may be thought that an alternative, seemingly more satisfactory approach to intelligibility is to derive the explicans directly from a set of premises that are themselves intelligible. But there is no direct way from such premises to the explicandum. The phenomena with which physicists are concerned and the premises by which they are to be explained have to be mediated by the language in which the phenomena are described. How else establish a relationship to the language of the premises? But a language, any language, whether verbal, symbolic, graphic, musical, or whatever, having a certain morphology, is itself a model of possible structures and processes. For example, a verbal description (in English and most other languages) of a phenomenon implicitly asserts that the phenomenon described has fairly distinct boundaries, instead of permeating the universe, and not to be abstracted from it. As M. B. Hesse points out (in *Models and Analogies in Science*, South Bend, IN: University of Notre Dame Press, 1966): "contrary to what some empiricist philosophers seem to have held, 'observation-descriptions' are not written on the face of events, to be described directly into language, but are already 'interpretations,' and the kind of interpretation depends on the framework of assumptions of a language community."

Today the quantum is recognized as the central principle of every branch of physics. However, in its many features of indeterminism, complementarity, the interference of probability amplitudes, it has sometimes seemed something strange, incomprehensibly imposed from outside on an unwilling world of physics. If it were fully understood, would its inevitability in the construction of the universe not stand out clearly for all the world to see? And could it not then be derived, along with all its mathematical superstructure, from some utterly simple first principle? Until we shall have arrived at this basic idea we can even say that we have not understood the first thing about the quantum story. If the situation poses a challenge and a question, is not the central role of the observer in quantum mechanics the most important clue we have in answering that question? Except it be that observership brings the universe into being, what other way is there to understand that clue?

However, this view raises other important questions:

If the act of observation, of registration, determines the phenomenon observed and registered, what, in the final analysis, is being observed and registered? W. Heisenberg answers (in *The Physicist's Conception of Nature*, New York: Harcourt, Brace, 1958): "The object of research is no longer nature itself, but man's investigation of nature. Here, again, man confronts himself alone." The investigator gets to know not nature, not reality, but only how the means and methods of investigation, and his own consciousness, are affected by the "objective situation" to which they are applied. Only indirectly, and on the basis of powerful assumptions, can inferences be drawn about the objective situation underlying the effects. The assumptions that have so far been employed, and the theories flowing from the assumptions have led, as we have seen, to fundamental paradoxes and mysteries, to unintelligibility.

If the totality of phenomena that constitute observed nature depends on acts of observation, if *esse est percipi*, as Berkeley argued, what was the universe before humankind, or any sentient being, appeared to perform the acts of observation and registration that determined phenomena? The answer to which more and more people, including some physicists, have been driven (see, in this connection, P. Davies, *God and the New Physics,* New York: Simon & Schuster, 1983) is essentially that supplied by Berkeley: that before there was Man, before any sentient beings, and wherever in the universe there is no sentience, there was, and still is, and will forever be, God, who observes and so determines the character of all that exists at every moment. Many people, including some physicists, find that a satisfactory answer. The difficulty, however, is "that such an assumption raises more problems than it solves" (P. Davies, *op. cit.*). The central point is that God, and the attributes commonly ascribed to Him (or Her) — His perfection, eternity, ubiquity, omniscience, omnipotence, etc. — are, by agreement of most theologians of most faiths, beyond any possibility of human comprehension, are utterly dependent on faith, as a supplement to, or substitute for, reason.

Whether or not one believes in a God (or gods), we must derive the character of His universe (if you will) from intelligible concepts about that character. If God himself is not intelligible, the assumption of deity, or faith in it, will not make the universe intelligible. Objective reality will remain as mysterious and incomprehensible as God himself is thought to be.

The condition that the concepts by means of which natural phenomena are to be explained must be about objective reality, not merely about the appearances registered by observation, "not about *observ* ables but about *be* ables, " has not been met either by quantum (or any other physical) theory, nor by the theological and esoteric doctrines to which many have fled in despair of finding understanding within science.

There is no good reason for such despair. While we can never gain direct knowledge of objective reality, phenomena can still be mapped or represented on models derived from what can be *reasonably* conjectured about the nature of that reality, not indeed its sense qualities (the way a God or an angel might experience it) but its structural or formal properties, presumably independent of our subjective way of interacting with and experiencing those properties.

I have emphasized "reasonably," for that is crucial. The first condition for the intelligibility of the models on which phenomena as observed and measured are to be mapped is that the premises from which the models are to be derived should be as reasonable, indeed as close to incontrovertability and self-evidence, as is humanly possible. This rules out the mechanical models, with their perfectly elastic balls, perfectly rigid rods, etc., which dominated explanation in physics until the twentieth century. With the triumph of Relativity Theory and especially of Quantum Theory, it is no longer reasonable to assume that nature is fundamentally like Lord Kelvin's mechanical models, or indeed like anything that we can visualize. What remains are only verbal, mathematical or other symbolic models. The first condition for intelligibility is accordingly that the phenomena to be explained be mapped on such models.

The second condition is that the premises from which models are to be derived must determine *specific* models, not all possible models, but only certain kinds, excluding all others. For otherwise there could be an infinity of possible models, and therefore explanations of phenomena, and all that would be determinate (if that) would be the phenomena modeled themselves. For that reason, the models may not be formulated in ordinary language, in that of the propositional calculus — contrary to Wittgenstein's dictum in the *Tractatus* that "proposition is a picture of reality" — or in the proliferating fields of mathematics. The reason is that the same phenomena can be represented in quite disparate mathematical systems, derived from different premises and ideas of order, as when Heisenberg represented quantum phenomena in terms of sets of matrices, Schröödinger in terms of wave equations, Dirac in terms of transformations in an abstract multi-dimensional space, and other theorists in still other ways. So, too, Relativity Theory can be expressed in terms of Riemannian geometry, but also (though less elegantly) in Euclidean terms or in terms of other geometric systems, incompatible with each other. No responsible theorist today holds — as Kant believed with respect to Euclidean geometry — that any given mathematical system, or set of propositions, or any other aggregation of verbal or symbolic utterances mirrors the structure of objective reality. Each is but one possible means, among many, of ordering data. The task of constructing a symbolism that does *uniquely* mirror the order of nature is still before us. (Whether or not the quest is vain is the central issue of this work.)

The third condition for the intelligibility of models is that the premises from which the models derive must afford the possibility of constructing models *adequate* to the task of representing *all* the phenomena to be explained. The models so derived must show how and why things happen the way they do, rather than in some other way; and in so showing, they must be able to resolve the paradoxes and mysteries that today haunt physicists.

The fourth condition for the intelligibility of models is that the underlying premises make it possible to show through modeling how the phenomena modeled *connect* with each other, how they *cause* each other. Otherwise the phenomena are isolated "atomic facts" and could well be different. The only intelligible causal connection is that of entailment, that is, necessary connection. Thus the fact than an ordinary die has six sides entails the probability of one-sixth that — if unloaded, fairly cast, and unaffected by wind or other influence — any given side will come up. So, too, the revolution of the earth around its axis and its orbiting of the sun entails the alternation of night and day and the succession of the seasons. Short of entailment, as David Hume demonstrated, attribution of causal connection depends on merely customary association of assumed cause and effect, however much such association may be enshrined in natural law and embodied in theory. For the question is: why these laws rather than possible others? Only if it can be shown that, in the nature of reality, they could not be other than they are, can the laws become intelligible.

The reasonableness of the premises from which models derive, the specificity of the models flowing from the premises, the adequacy of models to the phenomena modeled, and the connection by entailment of the phenomena modeled are the essential, indispensable conditions for the intelligibility of models. There are three additional conditions, less crucial, more readily dispensable.

The first of these *desiderata* is *depth*. The premises should lead to models that mirror the fundamental character of the phenomena to be explained: their ultimate constituents, the basic forces governing them, the essential unifying order of the universe. Otherwise the models and the explanations they support would not get to the bottom of things, and questions, mysteries, would remain, unanswerable, unresolvable, in terms of the models. Here, again, Einstein defined the ideal — the ambition to encompass the "totality of empirical fact" (*Gesamtheit des Erfahrungs-tatsachen*) in a simple, comprehensive, unified *Weltbild*. He fell under what the historian I. Berlin has called the "Ionian enchantment." And though this enchantment led the German *Naturphilosophen* of the nineteenth century and many other thinkers into error and confusion, it is an ideal still shared by most physicists, as the philosopher of science G. Holton notes in his contribution to *Scientific Explanation* (edited by A. F. Heath, *op. cit.*): "Today, as in Einstein's time, and indeed in that of his predecessors, the deepest aim of fundamental research is to achieve one logically unified and parsimoniously constructed system of thought that will provide the conceptual comprehension, as complete as humanly possible, of the scientifically accessible sense experiences in their full diversity."

This ideal is contrary to the anti-reductionism that is a strong tendency in modern thought, especially in the biological, behavioral, and social sciences. But even the anti-reductionists, while denying the attainability of the ideal, do not really dispute its desirability. Of course, should there be no ultimate constituents of reality, no truly basic forces, no unifying principles of universal order, this condition cannot validly be met.

The second desired condition is the *simplicity* of the premises from which the models flow, a condition that Descartes clearly defined (in *Rules for the Direction of the Mind,"* translated by F. P. LaFleur, Liberal Arts Press, 1961): "Method consists entirely in the order and arrangement of that to which the sharp vision of the mind must be directed in order to discover some truth. But we will follow such a method only if we lead complex and obscure propositions back step-by-step to the simpler ones and then try to ascend by the same steps from the insight of the very simplest propositions to the knowledge of all the others." Einstein repeatedly affirmed his faith in the attainability of this ideal, as in the lecture he delivered at Oxford University on June 10, 1933: "Our experience hitherto justifies us in believing that nature is the realization of the simplest conceivable mathematical ideas."

However, prevailing thought among physicists has tended to move away from Einstein's ideal, to doubt its realizability. Prigogene clearly expresses this view (*op. cit.*): "The only objects whose behavior is truly 'simple' exist in our own world, at the macroscopic level. Classical science carefully chose its objects from this intermediate range. The first objects singled out by Newton — falling bodies, the pendulum, planetary motions — were simple. We now know, however, that this simplicity is not the hallmark of the fundamental; it cannot be attributed to the rest of the world." Of course, the behavior of "falling bodies, the pendulum, planetary motions" are not truly simple. They were made simple in classical mechanics by abstraction from such complicating influences as wind resistance. And Prigogene is somewhat too positive about what "we now know." Yet the point he

makes does essentially reflect the position now tacitly held
by many, perhaps a majority, of physical scientists. But
while they may, perhaps justifiably, doubt the attainability
of the ideal, or even progress toward it, very few would reject the ideal itself. They would tend to agree that the prim-
itive concepts, the basic propositions and equations, of a
science must at least be simpler than the concepts, propositions, and equations derived from them and the phenomena they cover and purport to explain. If the basic ideas are
complex, how much more so must be the derivatives,
formed by the elaboration, combination, and permutation
of these ideas — until "complexity barriers" are reached,
perhaps very early in the development of a theory, beyond
the reach of our comprehension. Obviously, simplicity is relative, a matter of degree. While there is no absolute standard of simplicity, it is equally obvious that the simpler the
premises from which models derive, the more intelligible
will both premises and models tend to be.

Finally, there is a group of esthetic attributes, closely related to each other, and to depth and simplicity. Impossible
to define precisely — they include economy, elegance, uni-
formity, completeness, unity, symmetry, harmony, generality — they can perhaps be summed up in the word
"Beauty," to which physicists have been attracted in recent
years and which they have tried to exemplify in their own
theories and in those they admire and espouse. The more
beautiful the premises and the models they generate, the
more intelligible will they tend to be.

* * *

So much for the conditions — essentially those affirmed by
Einstein and others in the classical tradition of physics — to
be met. The rest of this book presents an effort to satisfy
them, and thus to approach Einstein's goal.

Chapter 2
The Postulate of Elementality

We must start with a reasonable conjecture about the objective nature of physical reality.

One conjecture that was first formulated in the Western world by Leucippus and other pre-Socratic philosophers, and that has cropped up repeatedly in various forms ever since then, is that the physical universe is composed of ultimate discrete constituents. (The alternative conjecture is that nature is fundamentally a continuum.) Today, with the triumph of quantum theory, atomism is in the ascendancy and indeed is not being seriously challenged. Everything is being atomized ("quantized" in current jargon): matter, energy, the fundamental forces, space (cf. the "bits" of Wheeler's pre-geometry), time (the hypothetical chronon). Leucippus' "void" itself is now conceived as a plenum of discrete virtual particles.

What Is Elemental?

The trouble is that we have today no clear idea of which , if any, of the particles or other discrete entities that have been observed or conjectured are truly elemental. (By "elemental," I mean, as do most physicists and philosophers of science, its ordinary dictionary definition: simple, uncomplicated, uncomposed, indivisible, irreducible.)

It has been abundantly clear, at least since the work of R. Hofstadter and his associates in the 1950s (cf. R. Hofstadter, *Nuclear and Nucleon Structures*, Reading, MA: Benjamin, 1963), that most of the particles that had until then been called elementary, so far from being truly elemental, were exceedingly complex (just as it became clear, toward the end of the nineteenth century, that atoms, once thought to be elemental, were just not so; and, before that, similarly with molecules). It was seen that most of "those objects which have been called 'elementary particles' have now to be considered complicated compound systems and should be calculated in some way from the underlying natural laws" (W. Heisenberg in *The Physicists' Conception of Nature*, edited by Mehra, *op. cit.*).

As a result of advances in experiment and analysis, the two families of hadrons, the nucleons (protons and neutrons) and other baryons, and the various kinds of mesons are no longer regarded as elemental, but instead as composed of quarks and gluons in various combinations. Most physicists today would claim elementality only for photons, leptons — electrons, muons, taus, electron neutrinos, muon neutrinos, tau neutronos, and their respective anti-particles (positrons, etc.), and quarks and their associated gluons. To these some would add the bits and chronons of space–time. The argument for their elementality rests mainly on their apparent pointlike character, their empirical irreducibility (at least so far), and the claimed potential of theoretically constructing all of physical reality — space–time, matter, energy, the fundamental forces — from them.

Against this argument, there are these considerations:

First, the great number and variety of the hypothesized ultimate constituents of physical reality: the twelve kinds of leptons; at least eighteen different kinds of quarks, and the same number of anti-quarks, making thirty-six in all, with maybe others still to come; at least eight kinds of gluons; three or more intermediate vector bosons, as well as photons, with wavelengths ranging from minute fractions of an angstrom (one hundredth-millionth of a centimeter) to many thousands of meters. And these are only the entities empirically observed or more or less convincingly inferred from observations. There are, besides, a host of hypothetical elementary particles: gravitons (the supposed carrier of the gravitional force), magnetic monopoles, tachyons (hypothesized to travel at velocities greater than that of light), instantons (solitons presumed to be confined to particular moments in time), superheavy bosons, Higgs particles (hypothesized to interact with the four gauge particles of the electroweak force), pseudoscalar particles, axions and other Goldstone bosons, lepto-quarks (hypothesized to be able to convert a quark to a lepton and vice versa), a grand assortment of elementary "virtual particles," which are supposed to be emitted by "real" (?) elementary particles and immediately reabsorbed by the same or other particles, or that supposedly exist "inside" elementary particles. In addition, during the last few years, theorists attempting to meld gravitation with the other fundamental forces have postulated a whole new, and large, family of "elementary" particles, including photinos, sleptons, squarks, gluinos, etc. — together possibly much greater in number of kinds than the now known particles and even of the others now conjectured. Other theorists have proposed a new family of heavy spin-1/2 particles called "techniquarks," subject to a new strong interaction called the "technicolor interaction." Such techniquarks would bind to form a plethora of techniparticles, such as techni-pions and techni-vector mesons. Given the technological dynamism that characterizes our civilization and the vast sums of money now being spent and the still larger sums now being appropriated or demanded, it is virtually certain that new, more powerful particle accelerators and ever more sensitive electronic detectors will reveal still more "elementary particles" and that the fertility of the scientific imagination, working on the new empirical findings, will generate whole new families of supposedly "elementary" particles. As R. Feynman says, with a hint of exasperation (in *QED: The Strange Theory of Light and Matter*, Princeton University Press, 1985): "As we go to higher and higher energies, nature seems to keep piling on these particles as if to drug us."

Such a proliferation of the observed or hypothesized "elementary particles" is less and less acceptable to increasing numbers of theorists, just as was the "particle zoo" that burgeoned wildly in the 1950s and 1960s, and that the quark model helped to tame. Surely, it is more and more widely and strongly felt, all these particles, in their multiplicity and variety, cannot all be elemental. Responding to this feeling, certain theorists have postulated the existence of particles more elementary than those so far detected, produced, or conjectured (sub-leptons, pre-quarks, preons, rishons, spinors, etc.). So far, there has been no evidence, nor convincing argument, for their existence. More significant is the thesis of the GUT theorists that, in the first microseconds of creation, before the primordial symmetry was, in stages, broken to produce the different forces, fields, and sub-atomic particles we have today, the elementary constituents of the nascent universe were all unified, so that, at the beginning, all was one. But while the electromagnetic and the weak forces have, provisionally, been unified in an electroweak force, with the particles carrying the forces (photons and intermediate vector bosons, respectively) being regarded as manifestations, resulting from broken symmetries, of the same particle, or family of particles, there is no evidence, nor convincing argument, relating the electroweak force to the strong nuclear force, let alone to the gravitational force. Moreover, the models advanced to unify the electro-magnetic and nuclear weak force, and their associated particles, are highly questionable, depending as they do on the hypothetical existence of Higgs bosons, supposed to interact with the particles of the electroweak force. The Higgs boson is today regarded as a mathematical fiction, about which (as about other aspects of the electroweak models) serious theoretical difficulties remain unresolved.

In sum, there are just too many kinds of particles called "elementary" to please theoretical physicists and philosophers of science; and the efforts to get beneath them to levels more elementary or, alternatively, to meld various kinds of particles into a smaller assortment of kinds have, in the main, been frustrated so far.

The second set of considerations militating against the claims for elementality for the particles now considered elemental concerns their mass — in some of the particles, relatively enormous masses. Thus the muon has 200 times the electron's mass and the tau has 3500 times the mass of the muon. So, too, certain kinds of quarks are thought to be hundreds and even thousands of times more massive than other kinds. But a mass is a magnitude. Quantum theory requires that mass-energy be quantized, which means that all particle masses must be expressed as a multiple of the elementary quantum of mass-energy. Thus, the mass–energy of an electron is a certain multiple of that quantum; the mass of muon is a certain multiple of the mass of an electron; and so on. But, by definition, that which is elemental must be uncomposed, indivisible, irreducible. The mass of an electron — and the more so of the more massive particles — can, in principle, even if not experimentally, be divided. It cannot therefore be elemental, in an important sense of the concept.

The third set of considerations against the claims to elementality of the present claimants concerns their structural characteristics: their possession of one or more of such attributes as charge, spin, frequency, "strangeness," and "charm." Even massless photons are characterized by frequencies, therefore structure. As S. Hawking remarked: "Each time we have extended our observations to smaller length scales and higher energies, we have discovered new levels of structure" (S. Hawking lecture, *op. cit.*): The fact that it has not been empirically possible to decompose leptons and hadrons (let alone the constituent quarks of the latter) does not, of course, prove that they are simple, rather than composite.

Finally, there are the profound theoretical difficulties with the standard model of sub-atomic structures. Treating particles, such as leptons and quarks, as dimensionless points meant the introduction of infinite quantities (e.g, the infinite self-energy and therefore mass of the electron) in the equations. But this was obviously absurd. Nature doesn't exhibit infinite physical quantities. To get around this difficulty, the procedure of renormalization was invented and deployed. This procedure simply eliminated the infinite terms. Though it was recognized as a "mathematical trick" it proved to have a limited effectiveness (and, therefore, presumably, validity) particularly in the elaboration of quantum electrodynamics by R. Feynman, J. Schwinger, S. Tomonaga, and others. But this still left many other domains of quantum theory "unrenormalizable." Moreover, even in the renormalizable fields, certain troublesome anomalies appeared. Viewed from a certain distance, a particle's mass or charge might have a certain value. Viewed at much shorter distances, as in a high-energy particle accelerator, the same particle would appear to have a mass or charge either substantially larger or substantially smaller than its long-distance value. Again, such anomalies were treated by application of a new branch of mathematics (developed mainly by B. B. Mandelbrot) called "fractal geometry," which has had a limited success in treating the anomalies.

However, dissatisfied with such "mathematical trickery," an increasing number of theoretical physicists, among whom M. Green, J. Schwarz, and D. Gross are today prominent, have devised a variety of theories in which the "elementary" particles are conceived as "strings" or "superstrings," either open or closed, instead of as dimensionless and structureless points. Their theories have excited wide interest, and work is proceeding on many fronts. Despite the success of these theories in treating the problems of re-normalization and of dimensional anomalies, and the fineness of their fit to many experimental phenomena, certain difficulties remain. Most of them require ten space-time dimensions (some more), all but four of which are "hidden." More significantly, nobody has yet formulated a general physical principle underlying the strings — why strings, rather than clumps, say, or other agglomerations? Why the ten or more dimensions? And so on. Because there have been no answer to such questions, most physicists, though fascinated by the new theories, are now inclined to regard them as still essentially "recreational."

More pertinent to my present argument: If strings or super-strings, whether open or closed, they are not points, but composed. Though the strings are hypothesized to be extremely short, in the order of 10^{-35} cm, they are *not* of zero dimension. They are supposed to have extension and to be composed of certain unknown and so far unimagined constituents. The questions inevitably arise: What are these constituents? What is the nature of the force or interaction that holds them together, that determines their dimensionality, their configurations, their role in the composition and the characterization of the empirically known sub-atomic particles? To such questions, thus far no answers.

* * *

The upshot of the discussion so far in this chapter is that, in view of the multiplicity and variety of the so-called elementary particles, of their massiveness, of their structural characteristics, of the theoretical difficulties in treating them as dimensionless and structureless points, it is no longer reasonable to think of the particles still called "elementary" as truly so. They are only sub-atomic (and that is how I shall henceforth refer to them). It is a fair presupposition that they are in some way composed of something truly elemental. But what is that something? And how can one go about finding answers, at least constructing plausible and useful models and theories? How can we go deeper?

Not empirically. As Heisenberg pointed out [in a lecture reprinted in *Physics Today* **29**(3), 1976]:

> How could one divide an elementary particle? Certainly not by using extreme force and very sharp tools. The only tools available are other elementary particles. Therefore, collisions between two elementary particles of very high energy would be the only process by which the particles would eventually be divided. Actually they *can* be divided in such processes, sometimes into very many fragments; but the fragments are again elementary particles, not any smaller pieces of them; the masses of these fragments resulting from the very large kinetic energy of the two colliding particles. In other words, the transmutation of energy into matter makes it possible that the fragments of elementary particles are again the same elementary particles.

Because the kinetic energies generated by the new particle accelerators are many times greater than those attainable a decade ago, the new particles now being created tend to be correspondingly more massive and, presumably at least, more highly structured than those previously known; and the particle generators now being built or planned will continue in the same direction.

Nor is it at all likely that the elements of the sub-atomic particles will be found in cosmic radiation or in the spontaneous decay of protons (should it occur). For to be observable, or to cause observables to occur, requires that the entity observed be a certain mass–energy packet; that it exhibit momentum, and also spin, charge, frequency, or other properties. Something utterly simple, without magnitude or structure, therefore truly elemental, can never be observed nor, by itself, cause the appearance of observables. The existence of elements, properly so called, can only be inferred from observation — or assumed.

It cannot be *definitively* inferred. For any number of hypothesized sub-particle entities and their presumed properties and behavior can be made to fit the observed findings at the particle level. Hence the recent proliferation of theories about the ultimate elements of the subatomic particles: preons, rishons, etc. As with the hidden-variable theories, any number of theories can be made to fit the observed phenomena. (The mathematician H. Poincaré pointed out, almost a century ago, that an infinity of curves can, in principle, be drawn through any set of points, however dense, representing discrete instrument readings.) Only some overriding consideration, some principle, or set of principles, of natural order, would make it possible to choose among possible entities (and the theories describing their properties and behavior), the one to be preferred to all the others. Otherwise we are left only with an array, potentially endless, of arbitrary and speculative fabrications.

But an ordering principle cannot arise from observation. It must, in Einstein's words, be "a free intellectual creation." "No ever so inclusive a collection of empirical facts can ever lead to the setting-up of such complicated equations [as those of General Relativity Theory]. A theory can be tested by experience, but there is no way from experience to the setting up of a theory. Equations of such complexity as are the equations of the gravitional field can be found only through the discovery of a logically simple mathematical condition which determines the equations completely or (at least) almost completely" (from Einstein's "Autobiographical notes," in *Albert Einstein: Philosopher-Scientist, op. cit.*) It is the purpose of this work to suggest such principles of natural order, such "a logically simple mathematical condition."

The First Premise

The first such principle is the *Postulate of Elementality*: that the constituent entities and processes of the reality we know: — matter, energy, the fundamental forces, space, time, whatever else there may be, without exception, in physical reality — are all composed of *elements* and *elemental events* (happenings of, to, and among elements), together the ultimate constituents of that reality and of the changes it undergoes. *Structures* conceived statically — though no truly static structures exist — are composed of elements. *Processes*, which change structures, are composed of elemental events. Elements are defined as utterly simple, accordingly uncomposed, indivisible, irreducible, without extension or mass or energy or other magnitude, and devoid of any intrinsic property other than their simplicity. Elemental events are similarly defined as utterly simple, therefore uncomposed, indivisible, unarticulated, instantaneous (that is, without duration).

From this postulate, four corollaries can immediately be inferred:

1. *Intrinsic Identity.*   Any element is *intrinsically* identical, and remains identical, with any other; and any elemental event is, and remains, *intrinsically* identical with any other; however much elements or elemental events may differ and change *extrinsically,* that is, with respect to their respective *histories* of relation to and interaction with each other, and only with respect to such histories. (The meaning of histories will be considered in the next chapter, and their significance through the rest of this book.)

For, being uncomposed, elements and elemental events cannot differ in composition. Being without magnitude, they cannot differ in magnitude. Being without intrinsic properties of any kind other than their simplicity, a property that is the same and unchanging for every element and elemental event, they cannot differ in any property. Being unlocated in space–time, but rather constituting it (as also matter, energy, force, and whatever else there may be) they cannot differ in space–time position. Only their histories of relations and interactions determine their locations with respect to each other; but they do so only extrinsically. Since (as we shall see under "dyadal interaction" below) any element interacts with only one other element at any one time and since all interactions are identical, elements cannot differ in the number or character of their interactions. (They can, however, differ *extrinsically* in the history of their interactions.) *Therefore,* every element is, and remains, intrinsically identical to every other; and every elemental event is, and remains, intrinsically identical to every other. They can differ only extrinsically.

2. *Action.*   Elemental events are all actions (the nature of which will be analyzed in the next chapter) of particular elements upon particular other elements.

For, since, by definition, elements are uncomposed and indivisible, therefore without internal constituents, dimensions, or structure, nothing can happen "within" an element. Accordingly, there cannot be self-action of an element. It follows that any elemental event must be external to each of the particular elements involved in the event. Since, by the Postulate of Elementality, all of physical reality is composed of elements, and of nothing else but elements, there can be nothing external to a given element except other elements. *Therefore,* all elemental events are actions of particular elements upon particular other elements.

3. *Availability.*   Every element has at least one other element, but not all other elements in the universe, available for interaction with it.

For, since, by the Postulate of Elementality, everything, including "space," is composed of elements and elemental events and of nothing else, nothing can intervene among elements and elemental events except other elements and elemental events. (That is, nothing but elements and elemental events can be or come to be between elements and between elemental events so as to separate them.) Given three elements (provisionally labeled *a, b,* and *c*) element *b* may separate *a* from *c*, but then *b*, which is itself an element, will be available to *a* and to *c*. Similarly, given three elemental events (provisionally labeled 1, 2, and 3) 2 may separate 1 and 3, but 2 is itself an elemental event, which will be available (that is, unseparated from, continuous with) both 1 and 3. *Therefore*, every element must always have one or more other elements available for interaction with it; every elemental event must be continuous with other elemental events.

So, too, since, by definition, an element is simple, therefore indivisible, it cannot extend over the entire universe, to be available for action, in the same instant, to all other or any other element in the universe. (The universe, as virtually everyone would agree, is extended, and an extension is, by definition, divisible; while an element is not.) A particular element may be unavailable at particular times to particular other elements, but it may have been available in the past, or may become available in the future, to particular other elements anywhere in the universe.

Therefore, while some elements are always available for interaction with any particular element, not all elements are, at any particular time, available for interaction with each other. Particular elements may, at any specified time, either be available to each other or not so be because of intervening elements; they may remain or not remain available to each other, depending on the absence or presence of intervening elements. An elemental event is continuous with its immediately preceding, and its immediately following, elemental event, but it is not continuous with other elemental events, or with sequences of such events.

4. *Dyadal Interaction.*   Any element can act upon *only one* other element at any one time, and must so act, and the element acted upon must, at the same instant, react upon the acting element, and no other, so that all elemental events are instantaneous dyadal interactions.

For, as to the necessity of action and interaction at each instant; by the Postulate of Elementality, time is composed of elemental events, each instantaneous. Without events, there would be no time. With nothing happening to an element, no time would pass for it. Every instant of its existence, without exception, must be occupied by, is constituted by, an action by and upon it. *Therefore*, at every instant, every element must act upon some other, and the element acted upon must, at the same instant, react upon the acting element. The action and the reaction are of the same character, perfectly symmetrical, accordingly an interaction.

As to the limitation of each interaction to a single pair of elements (an element dyad),* for an element to interact with more than one other element at any one moment would require that it be composite — one part or aspect interacting with some other element, another part or aspect with a fourth element, possibly still another part or aspect with a fifth, and so on. But, by definition, elements are uncomposed and indivisible, without parts or aspects, without distinction of surface and interior, or other attributes of structure, the components of which can be separately engaged. *Therefore*, elements being utterly simple, and elemental events also being utterly simple, only pairs of elements, and never more than pairs, may interact with each other at any given moment.

So, too, were two interactions of the same couple of elements to occur in the same instant, the instant would have to be composite, one part or aspect occupied by one interaction, another by another. But an instant (which is determined by an elemental event) is, by definition, instantaneous, without duration, therefore indivisible. Were an elemental interaction to take more than one instant, it would have to be composite — one part of the interaction

---

*I shall employ the term dyad in reference to two interacting elements considered as a unit. I shall use the term "pair" or "couple" when I consider them individually, as in saying "a pair of elements interacts, so becoming a dyad." I use instant and moment synonymously, therefore interchangeably.

occupying one instant, another part another instant, and so on. But element interactions, which are the elemental events, are, by the Postulate of Elementality, simple, therefore uncomposed. Though every event must involve two elements (but no more than two) — there being no self-action of elements — each acting upon the other, the interaction is simple, uncomposed, indivisible, irreducible. An action with reference to (from the "viewpoint" of) any one element is a reaction relative to the other; but it is the same action, one interaction, not two. It is not a matter of an action causing a reaction — an element cannot be caused to react; it must, at every instant, interact. The extrinsic identity of the *particular* other element with which it interacts at any given instant is determined by their respective histories — but *some* reaction at every instant is inescapable and, in that sense, uncaused, therefore not really a *re*action. Action and reaction constitute the same event, as seen from two different perspectives, that of each of the two elements interacting.

Therefore, every instant in an element's history is determined by an interaction with some other element, never more than one. Every interaction is instantaneous, that is, it occupies only one instant, never less nor more than one instant.

Alternatives to the Postulate

How well does the Postulate of Elementality fulfill the conditions for intelligibility stated in Chapter 1?

First, as to the most crucial condition, that of reasonableness, most readers would, I believe, agree, that it is a reasonable conjecture. But it is far from incontrovertible, let alone self-evident. Other views about the ultimate constituents of physical reality (including the view that there are no ultimate constituents) can be, and have been, persuasively argued. Among these views (a list by no means exhaustive):

1. That of the strict empiricists, in the tradition of the Austrian physicist and philosopher Ernst Mach who urged (as did some of his followers among the logical positivists who flourished in the 1930s, 1940s, and 1950s) that science must concern itself solely with "sense impressions" and "pointer readings" and their proper ordering, and who accordingly rejected the reality of atoms (let alone of subatomic particles) as not being objects of direct observation and measurement, and therefore purely, and objectionably, speculative.

2. That there is no objective world "out there," and instead that the world (or worlds) we know is constituted by the ways we think of it, the paradigms that direct and structure our thinking, the languages we employ to express our thoughts. [This view was given its classic formulation by Kant, and in our own day has been most trenchantly argued by T. Kuhn in *The Structure of Scientific Revolutions*, 2nd edition (University of Chicago Press, 1970); by N. Goodman in *Ways of Worldmaking* (Harvard University Press, 1978) and *Of Mind and Other Matters* (Harvard University Press, 1984); and by M. Foucault in *Les Mots et les Choses* (Paris: Gallimard, 1966), and in many other works.]

3. That the sub-atomic particles, as also energy, space, and time, are divisible, at least in thought, without end. This would mean that the known or conjectured sub-atomic particles are composed of sub-particles, and these, in turn, of sub-sub-particles, and so on inexhaustibly; and similarly for interactions among particles, for energy, space, time, and whatever else there might be. (Certain Soviet physicists and philosophers of science have urged such ideas. See, for example, I. V. Kuznetsov and M. E. Omel'yanovskii, eds., *Phil-*

osophical Problems of Elementary Particle Physics (New York: Daniel Day, 1966.)

4. That there are different kinds, possibly an infinity of different kinds, of constituent elements, for baryons, leptons, space–time, etc.; and that it is accordingly hopeless to attempt to conceptualize objective reality, to make it intelligible, by postulating only one kind.

5. That the whole universe is "implicated" in each of its elements, making the presumed elements, in some sense, coextensive with the universe — an idea developed, somewhat vaguely, not to say mystically, by D. Bohm in his *Wholeness and the Implicative Order* (London: Routledge & Kegan Paul, 1980).

6. That the questions: "What are the sub-atomic particles composed of? What are their elements?" have no real meaning, because what is fundamental in nature is process. In this view, "the world is a net of quantum processes represented by a finite-dimensional vertex rather than particle fields interacting in space–time. Further, in our process view of nature, the elementary constituents of nature are not material particles having masses and positions, but monads, elementary quantum processes" (from D. Finkelstien and G. McCullum, in *Quantum Theory and The Structure of Time and Space,* L. Castell, M. Drieschner, and C. F. von Weizsäcker, editors, Munich: Carl Hanser Verlag, 1975).

7. That what is fundamental to nature is not any kind of elemental particle, but rather symmetries and other mathematical relations. This view, of which Eddington, Jeans, and Heisenberg were among the most prominent early exponents among physicists, has found increasing favor among theorists. This has certain affinities to Plato's conception of reality as composed of ideas, essences, forms; and the Hegelian notion that reality is constituted by the dialectical opposition of ideas (e.g., "being" versus "not-being" resulting in "becoming").

8. That nature, physical reality, is *not* ultimately simple.

(Compare the view, already cited, of Prigogene.) Indeed, some physicists and philosophers now profess to believe not only that there is no first, simple level in the hierarchy of natural organizations, but that complexity increases, rather than diminishes, as we *descend* to ever lower levels.

One could go on. But the views indicated appear to be most worthy of note. None of them can be decisively refuted. Any of them may be "true." But they all lead to dead ends.

The first view, that of the strict empiricists, was largely discredited by the work of E. Rutherford and others who, early in this century, established, by impeccable experiments, not only the atomicity of matter but that atoms themselves are composed of sub-atomic particles. The Mach position has now been almost universally abandoned. It doesn't even matter very much to physicists today that particular particles (e.g., gluons) cannot be empirically detected. It suffices that their existence is required by the prevailing standard model and is comformable to it.

The second view, that of the constructivists, is, to my mind, incontrovertible. The worlds we know *are* constituted by the infinitely varied ways we individually perceive and conceive the world, But it is also true, I believe, that there are objective realities that remain invariant through all cognitive and linguistic transformations.

The third view, that of infinite divisibility, would mean that basic questions about physical reality would fall, inescapably, into the domain of metaphysical speculation and remain lost forever in the labyrinthine paradoxes of the infinite.

The fourth view, that of many different kinds of elements, lacks simplicity as well as empirical grounding. It would appear to be intellectually unmanageable, therefore sterile.

The fifth view, that of the implicative order, is unclear to me. Therefore I cannot responsibly comment on it, except to say that it has interested only a small coterie of physicists and philosophers of science, and that it doesn't appear to hold promise of anything significant.

The sixth view, that of quantum processes as being fundamental, is, like the fifth view, beyond my power of full comprehension. Again, it has had little influence and seems unlikely to generate much.

The seventh view, that what is fundamental in nature are not elements, or any other physical entities or processes, but purely mathematical relations, would mean that there is no substance in the universe, no matter, no energy, no forces, no space, no time, but only abstract ideas in the minds of men or of God (cf. Jeans' famous dictum: "God is a geometer"). But the ideas do not coincide. There are many different, not necessarily compatible, views of the fundamental symmetries and of breaks in symmetry. More to the point: A relationship is the way things stand with respect to each other. Without things, objects of thought, whatever their nature, to be related, there can be no relationships. (If there is a Cheshire cat's grin, there must be a Cheshire cat.) The ancestral Pythagorean, Platonic, and Hegelian views are of interest today mainly to historians of philosophy.

The eighth view, that nature is not ultimately simple, has, to my mind, the most to commend it. Yet it is essentially defeatist. If complexity increases, rather than diminishes, as we descend to lower levels of the natural hierarchy of order, how can we ever make sense of nature? As we go deeper into it, we would encounter ever more formidable complexity barriers and, at its center or bottom level, an impenetrable riddle. It is true that one doesn't always have to reckon with complexities at lower levels since only the resultants, the net effects, of lower level processes matter. It is this consideration that largely validates the ignoring of micro-physics by geophysicists and observational astronomers (though not by astro-physicists), of bio-chemistry by zoologists, of physiology by psychologists, and so on. *If* the force of gravity is independent of the chemical composition of the bodies accelerating toward each other — which, as noted above, is now being questioned — the complexities of chemical analysis can be ignored by those concerned with gravitation. So, too, people caught up in a panic or a mob act in a manner largely independent of their individual anatomical or physiological states, which can accordingly be ignored in studies of such behavior. It is not that the gravitational interaction of massive bodies is simpler than the chemical composition and structural features of these bodies; nor is it that psychological states are simpler than the underlying neurological, hormonal, and other physiological processes. They are *made* simpler, merely by ignoring or slighting the underlying complexities, by abstracting from them what is regarded as essential. The strategy works *if* the effects of the complexities are indeed negligible. Whatever the accommodations of nature (e.g., the "law of large numbers") that allow us to ignore complexity, or the complexity barriers that prevent fully and precisely reckoning with it, it remains true (at least to my thinking) that each higher level of organization is progressively more complex than the levels below it, because it encompasses and often complicates these complexities. Conversely, each lower level is simpler than those above, and the lowest, the elemental, level (assuming there is one), whatever its complexities, is the simplest of all.

The only escape that I can see from such dead ends is to posit elementality as I have defined it — acknowledging that it is only an assumption and, as such, not requiring demonstration — and then to find out what it leads or contributes to. If nothing much comes of it, then it too will prove to be a dead end like the alternative views I have presented.

So much for the first condition of intelligibility, that of reasonableness. What of the other conditions stated?

The Postulate has depth, since it refers to the ultimate constituents of all of nature. And it is simple — essentially the idea that all the complexities of our experience and of

our valid theorizing can be analyzed into simple, and ultimately, the utterly simple.*

The other conditions for intelligibility — specificity, adequacy, connectivity, and beauty — cannot be met by the
Postulate of Elementarity (and its corollaries) alone. Other
premises are needed, and a formalism derived from the
premises, and models constructed with the formalism, and
explanations based on the models, before the test of intelligibility can be meaningfully applied. That is the task for
the rest of this book.

*Indeed so simple and intuitively clear is the basic notion advanced that
similar ideas have occurred to and been urged by a great many people,
both naive and sophisticated, starting with certain ideas vaguely expressed in surviving Egyptian, Vedic, and Taoist writings, continuing
with the pre-Socratic atomists, and then the Roman, medieval, and re-
naissance alchemists, who were obsessed with the notion that there exist
certain elemental particles of which everything in the universe is composed. Such ideas were simplified by the Jesuit mathematician R. G. Bos-
covich, who rejected the concept then prevalent among alchemists and
other natural philosophers of an assortment of different kinds of natural
particles and held instead that the fundamental particles of matter were
all identical point-centers and that differences in appearance arose from
differences in their spatial-configurations. C. F. von Weizsäcker, in *The
Unity of Nature* (New York: Farrar Straus & Giroux, 1980) advanced the
Postulate of Ultimate Objects, that all objects consist of ultimate objects
(*urobjecte*) in a Hilbertian space; and the Postulate of Symmetry: that
none of the *urobjecte* and none of their states is objectively distinguishable
from any other. Most recently, as already noted, there has been a profu-
sion of theories about preons, rishons, and other presumably elemental
constituents of the sub-atomic particles. Nothing much has so far come
from any of these ventures. One reason is that the ideas are not simple
enough — von Weizsäcker, for example, complicates the two postulates
noted above with a number of others much less elegant. Another reason
is that they do not conform fully with the standard model of sub-atomic
structures, except by artificial force-fitting, which is also the weakness of
the hidden variable theories. Third is a basic inadequacy of the variant
formulations of the Postulate of Elementality — including my own. Additional premises are needed, which have not so far been satisfactorily
supplied, a deficiency I shall try to remedy.

Chapter 3
The Principle, the Rule, and the Formalism

The first supplement to the Postulate of Elementality is The *Principle of Otherness*: Given any two elements, available for interaction with each other, each is not the other, even though they are intrinsically identical. Given any two elemental interactions, each is not the other, even though they are intrinsically identical. From the supposition that there are two elements, two interactions, indeed two of anything, it follows that there must be one and another, one not being the other; for, were one the other, there would be only one, not two. This principle applies to all elements and to all their actual or possible interactions.*

---

*This Principle appears to conflict with Leibnitz' Principle of the Identity of Indiscernibles, which affirms that it is impossible that there should exist things that differ *sole numero*, that is, only because they are two or any other number, being otherwise identical. "To suppose two things indiscernible is to suppose that same thing under different names" (fourth letter to Clarke). I agree that the same thing may *not* have two different names (as I shall later have occasion to stress) nor may two identical things have nonidentical names; but they can, indeed must, have *other* names, which though clearly indicative of *present* identity, also indicate a different *history* of elemental interactions. Identical elements are indis-

The Principle of Otherness fulfills the first of the essential conditions for intelligibility. That there are ultimate elements and elemental events is assumed, but if there are, the Principle must apply to them. This is not merely reasonable, but incontrovertible, since it is an analytical proposition, therefore a tautology, the truth of which follows of necessity from the definition of its terms. The meaning of "two" is one and another; the meaning of "three" is one and another and still another; and so on for all the integers.*

But the Principle of Otherness is not, in itself, adequate for intelligibility; nor does its conjunction with the Postulate of Elementality and its corollaries make it so. Indeed, in itself, it is no more than the most trivial and vacuous of tau-

---

cernible — except in their histories of interactions, which Leibnitz did not take into account. In each interaction, the interacting elements are intrinsically identical, but one may have a history of repeatedly interacting with only one other element, while the other's history is that of interaction with a number of other elements. While intrinsically identical, they differ extrinsically, not in their *present* interaction, always the same for all elements, but in the number and order of the other elements with which they had previously interacted and in the respective histories of each of these elements.

*There are other definitions of integers, e.g., those of G. Peano and of G. Frege. Being definitions, therefore analytical, they are incontrovertible. However, the definition I propose is not only simpler than possible alternatives but (as I shall try to show) has more significant consequences for the understanding of objective reality.

While the Principle of Otherness is an analytical proposition, therefore true a priori, it is not a *synthetic* a priori proposition, in Kant's sense, since the *existence* of the elements and elemental events numbered is not self-evident a priori, but has to be assumed, in the Postulate of Elementality. Indeed, the existence of anything at all, of the universe itself, is not a priori certain, nor can it be decisively proved. What we perceive, what we think we perceive, may be projected into our minds by a "demon," as Descartes argued.

tologies. The only way it can be given substance and significance is so to specify its formulation and the formalism generated by that formulation, that inferences can validly be drawn from such specification (the way inferences can be drawn from specific formulation of what we know about the composition and spatial structure of a chemical compound, or about the design of an electronic circuit, or about the boundaries, contours, and other characteristics of a terrain) for the objective reality so modeled — the validity of the inferences depending, in each case, on the correctness and precision of the formulations.

For inferences to be validly drawn, the formulation specified must be a *uniquely necessary* representation of what is known or conjectured or assumed about whatever is to be represented. Only so can the condition of specificity be met. While there may be many different ways, indeed in principle an infinite number of possible ways, of representation — e.g., a map may be of any size one chooses, drawn on an unlimited variety of paper, metal, stone, or other surfaces variously textured and shaped, with all conceivable sorts of inks, pigments, incisions, etc. — all the ways must be *structurally* identical, the structure shown by each corresponding (with the desired or feasible degree of accuracy) to the terrain mapped. Should representations structurally different be permissible, different, perhaps mutually contradictory, inferences may validly be derived from structurally different mappings of a terrain, formulas of a chemical compound, designs of an electronic circuit, scorings of a musical composition, or whatever else may be represented. Conversely, *if* we can uniquely represent what is known, conjectured, or assumed about any given object or process, we can deduce valid inferences from the representation of what is known, conjectured, or assumed about the object or process represented. This is the significance of the discovery and confirmation of the double-helix structure of the DNA macromolecule, and its structurally unique modeling (with the degree of precision permitted by the state of our knowledge and understanding).*

* * *

Accordingly, for the required uniquely necessary formulation of the elements and elemental interactions I have postulated and of the Principle of Otherness that governs them, I need an Exclusion Rule, or set of such rules, implicit in what I have assumed and inferred, which, by excluding all alternative formulations, lead to a uniquely necessary, therefore fully determinate, specification of what I have assumed and inferred.

Exclusion rules are among the most important ordering concepts in the sciences. Chemistry, as a scientific discipline, originated with Lavoisier's assertion of the Law of Definite Proportions, which requires that every chemical compound be represented as composed of the same set of atoms in the same proportions by weight. Implicit in this law is the exclusion rule that we may not represent a particular compound as composed of different assortments of atoms in indefinite or varying proportions. At the heart of thermodynamics is its Second Law, which prohibits representing the entropy of an isolated system as decreasing. Relativity Theory is largely grounded on the rejection of an absolute or preferred space–time frame. Implicit in this recognition is the rule prohibiting the formulation of the laws of nature with reference to some privileged space–time frame or in such ways that the laws do not remain invariant through all continu-

---

*Of course, to the extent that experts differ about DNA's structure, e.g., about the precise sequence of nucleotides in the DNA of a particular species of plant or animal, and the precise role of DNA in morphogenesis — and differences do, of course, persist — there will be different representations of DNA's structure and operation, and different inferences will follow.

ous transformations of given space–time frames. The central concept of Quantum Theory is the Uncertainty Principle, which affirms the impossibility of jointly determining, with perfect accuracy, the position and momentum of a subatomic particle. From this follows the rule that we may not attempt to co-express them with perfect precision. Basic to the quantum theory of atomic structures is another exclusion rule, the Pauli *verbot*, which declares that "there can never exist two or more equivalent electrons in an atom which, in strong fields, agree in all quantum numbers." Implicit in a conservation law is the rule that, we may not, in describing a physical structure or process governed by the conservation law proposed, affirm or represent either an increase or diminution, however slight, of the properties said to be conserved. One could cite many other such exclusion rules.

The exclusion rule, or rules, I seek, which would lead to a uniquely necessary formulation of the Postulate of Elementality (and its corollaries) and the Principle of Otherness, may not be an arbitrary rule, or set of rules. They must be implicit in the Postulate and Principle stated, just as the exclusion rules cited above were all implicit in the central concepts and principles of their respective theories.

I submit that implicit in the Postulate of Elementality (and its corollaries) and the Principle of Otherness is the *Rule of Unambiguous Representation*: If elements and their interactions, and the combinations and permutations of elements, are just what they are or are assumed to be and not other than they are or are assumed to be, each not being any other, and if they are to be represented just the way they are or are assumed to be and in no other way, then we must give elements and their interactions, and the combinations and permutations of elements and interactions, unambiguous representation, that is, representations that are definite, precise, complete, determinate, uniquely necessary (only one way and no other) — each element and each in-

teraction only in its own proper way, and each other element and interaction in other ways proper for them and so,
too, for their combinations and permutations, so that there
is never, in principle, the possibility of mistaking a given
element or interaction or any combination or permutation
of elements and interactions for any other, or that would
permit them to be understood as other than they are.

From the Rule of Unambiguous Representation, together with the Postulate of Elementality and its corollaries,
and the Principle of Otherness, the following sub-rules
derive:

1. *Sub-Rule of Dyadal Representation.*   By the Postulate
of Elementality and the corollaries of Identity and Dyadal
Interaction, we may only represent elements and elemental
interactions, and never any other entity or event, the elements all being intrinsically identical with each other and
the elemental interactions also being intrinsically identical
with each other; and the interactions represented must each
involve two elements, neither fewer nor more than two.

2. *Sub-Rule of Representing Histories.*   Though intrinsically identical, elements differ extrinsically, with respect to
the particular other elements with which they severally interact and the patterns exhibited by sequences of interactions, as determined by the mutual availability and separa-
tion of elements and by how these change over time (that
is, over the course of a succession of interactions). The particular other elements with which given elements interact,
and the specific order of their interactions constitute the
histories of the elements involved. It is only in the history
of their interactions — not without significance, as we shall
see — that elements differ. (It is this extrinsic *non*-identity
that gives the Principle of Otherness its substance. Otherwise elements and their interactions could not be individually represented. All we could predicate of aggregations of
elements is their number and that they interact in pairs.) It

follows that the extrinsic identity of each particular element, which means the history of the interactions that determines its extrinsic identity, must be unambiguously, therefore uniquely, represented — uniquely because different representations could be variously understood; the extrinsic identity of each particular interaction, which means the identity of the couple of elements participating in it, must also be unambiguously, therefore uniquely, represented.

3. *Sub-Rule of Unique Representation.* It follows from subrule 2 that no expression in the formalism may represent more than one element state, constituted by its history of interactions. Conversely no element state may be represented by two or more expressions.*

---

*It may be noted that this rule defines the conditions for perfectly unambiguous discourse, not only about elements and their interactions, but about anything at all. Thus. F. P. Ramsey (in *The Foundations of Mathematics*, New York: Harcourt, Brace, 1931): "In a perfect language, each thing would have its own name." Similarly, B. Russell, in his preface to Wittgenstein's *Tractatus* (*op. cit.*): "The first requisite of an ideal language would be that there should be one name for every simple, and never the same name for two different simples." No such perfect or ideal language has ever been constructed. Every language, whether natural or artificial, employs class designations and other abstractions applicable to more than one, and indeed a potentially unlimited number of, individual entities or happenings or whatever. This is also true in *using* the formalism about to be presented to construct models for physical structures and processes. Beyond the representation of the very shortest element histories and of the very smallest element aggregations, the notations become so long and intricate, in a little while so unwieldy, that we must deal with them indirectly, by characterizing rather than actually representing the elements and interactions, and their aggregations. It thus becomes necessary, or at least expedient, to employ a meta-language, in which the same terms are used to refer to extrinsically diverse elements and interactions and in which different terms may refer to the same element or interaction. But first we must construct the rudiments of a formalism based on the strictest possible enforcement of the rules here stated, and draw the implications of such a construction.

4. *Constraint of Sufficient Reason.* How we represent elements and their interactions, and the combinations or permutations of such elements and interactions, and what we may say about any of them must be limited to what can be validly inferred from the uniquely necessary formulation of what is implied by the Postulate of Elementality and its corollaries, the Principle of Otherness, and the Rule of Unambiguous Representation and its sub-rules. Nothing may be assumed to exist or occur or to become more or less probable without sufficient grounding in the formalism and the premises from which the formalism derives.

While the Rule of Unambiguous Representation and its sub-rules are *not* necessary consequences of the Postulate of Elementality and the Principle of Otherness — no rule, which by its nature is normative rather than cognitive, can be deduced by necessity from cognitive propositions — they are strictly in keeping with the Postulate of Elementality and the Principle of Otherness. They are imposed on any effort to realize and apply these premises in representing elements and their interactions. I submit that no alternative rules would do.

The Formalism

Two characters — indeed two marks, sounds, or signals of any kind, as long as they are clearly distinguishable — will suffice to implement the rules just stated. Therefore, no more than two may be employed. This is more than the enforcing of Ockham's Razor: "*Entia non sunt multiplicanda praetor necessitam.*" More fundamentally, were three (or more) characters to be employed, when two would suffice, there would be more than one way — the two-character way, the three-character way or ways, etc. — of representing a particular element or interaction. But this would be contrary to the sub-rule of Unique Representation, which prohibits alternative representations of the same element or the same interaction.*

From the endless possibilities I have arbitrarily chosen, for my notation, 0 for the element sign, the operant, and / for the interaction sign, the operator. (As already emphasized, any other pair of distinguishable signals could just as well have been selected.) Neither of these signs is to be regarded as, in itself, representing either element or interaction. It is only as they are legitimately combined and per-

---

*With one character, or other signal, nothing can be constructed, only unpatterned repetitions of the same signal. With two, everything can be constructed — indeed, too much.

We know from computational and information theories, and with actual experience with computer and telecommunication systems, that any two distinguishable signals — 0 and 1, the presence or absence of an electric current or of a static electric charge, a polarized magnetic core or one depolarized, or any other pair of contrasting signals one might select, provided they can be electronically manipulated or be translated into signals that can so be — quite suffice for the representation of anything one may want, and their transformation in any desired way: verbal or mathematical constructions, musical and other sound patterns, pictures, diagrams, or other graphic material (in four dimensions and in full color, if desired), any specifiable object, state, or process, any algorithmic or heuristic procedure, any randomizing procedure or procedure for selecting among the transforms randomly generated. The only requirements (in addition to the distinguishability of the two signals employed) is that the instructions determining the combination and permutation of the signals be fully and precisely specified, and that there be enough computer throughput and communication channel capacity and enough time to accomplish given tasks. The limitless potentials of two-signal-based computation and communication prompted J. von Neumann, when asked by a little old lady whether there was anything a computer could not do, to answer: "Madam, if you can tell me *exactly* what it is that computers cannot do, I may be able to give you a computer that can do it' (As reported by D. Gabor in *Innovations*, London: Oxford University Press, 1970.) My emphasis on "exactly" — for that is, of course, the crux of the matter. The inability to specify given tasks exactly is what limits, and may forever limit, what computers can actually do.

In this spirit, P. W. Atkins recently wrote (in *The Creation*, Oxford and San Francisco: W. H. Freeman, 1981): "At heart the basis of the

muted that the resulting expressions can properly be said to represent possible elements or interactions.

Given now two elements, *provisionally* designated 0 and 00 respectively, and their mutual interaction at one instant, the elements and their interactions must be represented as shown below, and only so:

/0/00/   /00/0/

(Their relative position on the page and the space that separates them are not part of the formalism, and have no significance for it, any more than do the typeface, the color of the ink, the texture of the paper used. So, too, the expressions could just as well be written from right to left instead

---

universe must be as simple as the difference symbolized by 1 and − 1, or by yes and no, or (more prosaically) by true and false. The fundamental building blocks of the whole of creation must have this simple binary form. Nothing simpler has properties. Only the difference symbolized by 1 and − 1, by one and not-one, or point or no point, is sufficiently simple to be creatable, but rich enough when sufficiently concatenated (as in mathematics and in logic) to lead to properties. At root the universe is a dust of binary forms." Similarly, E. Fredkin has still more recently proposed that the whole universe, in its infinite variety and variability, is essentially a computer composed of the cosmic equivalents of yes and no, on and off, 0 and 1.

However, computerization's very fecundity, the absence of constraints upon it — except those inherent in the physical limitations of any mechanical contrivance and those arbitrarily imposed on it — is the trouble. It is possible to represent the wildest fantasies, even physical impossibilities (e.g., diminishing entropy in an isolated system or violations of gravity) in computer languages and programs, in full accord with their formation and transformation rules. The formalism here presented differs from the languages and programs of computation, telecommunication, and other modes of signal generation and manipulation in the constraints composed by the Rule of Unambiguous Representation and its sub-rules. My purpose is *to delimit the possibilities* of signal generation and manipulation to those strictly conformable to the Postulate of Elementality and the Principle of Otherness, and accordingly (presumably at least) to the nature of physical reality.

of left to right, vertically, diagonally, or in any arbitrarily-chosen slope, as well as horizontally. None of these attributes are significant aspects of the formalism.) Each of these expressions is an *element name*. It represents the element's identity at the moment of interaction, as determined by the interaction immediately represented and the known history of interactions in which it had previously engaged. The expressions *may* be read — though not altogether properly, since the operand alone is *not* an element name — as "' element 0' interacts with 'element 00'; 'element 00' interacts with 'element 0.'" (Demonstrating the unique necessity of this formulation and those about to follow will be attempted later in this chapter.)

To represent a second mutual interaction of these two elements, we must juxtapose the expressions representing the first interaction and add a stroke (the operator) at the end of each of the two new expressions, thus:

$$/0/00//00/0// \qquad /00/0//0/00//$$

The resulting element names represent the *states* of the given elements at the instant of the given second interaction — "state" being defined as the resultant of all known previous interactions, *in unbroken sequence,* of an element up to and including the current interaction. The expressions may be read (with the same qualification as above): "'element 0,' which *had* interacted with 'element 00' *now* interacts with 'element 00,' which *had* interacted with 'element 0'"; and similarly for the reciprocal interaction.

Let us now assume that, at the next moment, two other elements, provisionally designated '000' and '0000', respectively, become available for mutual interaction with the first two, creating a *set* of four elements — "set" being defined as any number of elements and their interaction or of representations of elements and interactions, selected for discussion in a particular context. (Some of the many different kinds of sets will be described below.) There are now three possible sets of representations of the third mutual interactions in this sequence, and *only* three. Each of the three sets is to be regarded as equally probable, since no reason is here given to the contrary, and by the Constraint of Sufficient Reason no differences of probability may be ascribed to occurrences without explicitly stated or shown sufficient reason. The first possible set is to be represented thus:

/0/00//00/0///00/0//0/00///
/00/0//0/00///0/00//00/0///
/000/0000//0000/000///0000/000//000/0000///
/0000/000//000/0000///000/0000//0000/000///

As is evident from inspection, the name of the first element in its second state '/0/00//00/0//' is juxtaposed to that of the second element '/00/0//0/00//' in its second state, a stroke being added at the end of the new expression; and the reciprocal juxtaposition is also made; while the name of the third element in its second state '/000/0000//0000/000//' is juxtaposed to the name of the fourth element in its second state '/0000/000/000//0000//,' and the reciprocal juxtaposition is also effected. In each case, a stroke is added at the end of the expression.* In the second set of possibilities, the names of elements 1 and 3 and 2 and 4 are juxtaposed in the manner indicated. In the third set, the element names of 1 and 4 and of 2 and 3 are juxtaposed. At each

---

*The added stroke is arbitrarily placed at the end rather than at the beginning of each merged expression, therefore the notation /0/00//00/0// instead of //0/00//00/0/. However, the added stroke could just as well have been placed at the beginning of the expression. The two expressions are structurally equivalent, because the difference doesn't affect any other feature of the formalism, just as the structure of an English word or an English sentence would be unaffected were it written from right to left in Hebrew or Arabic fashion, rather than left to right, or vertically, in the classic Chinese and Japanese way, rather than horizontally. A characteristic of construction is structural only if a change in that characteristic would also alter other characteristics; if, that is, it is an aspect of an interdependent system of construction features. All other features are non-structural, and can accordingly be decided upon arbitrarily.

succeeding moment of interaction, three new sets of possibilities would arise.

Should, at the fourth moment, two more elements become available for interaction with the first four, there would be 15 possible sets of interactions among the six given elements — again, each set being equally probable, there being, *as yet,* no reason to differentiate among probabilities. With each succeeding instant, there would be 15 new possibilities. As the number of elements mutually available for interaction increases, the number of possibilities also increases, though, of course, at a much greater rate, deducible from combinatorial analysis.

There is only one kind of operation in the formalism: the juxtaposing of pairs (never more than pairs) of equal length (a condition to be justified below) and the addition of a single stroke (never more than one) to the end (or beginning) of each merged expression. The number of *possible* juxtapositions of element names can be derived from combinatorial analysis; the *probability* to be ascribed to each of the possibilities depends on the special character (soon to be described) of the set of which given elements are members. But the number of resulting element names neither increases nor diminishes. Elements may be introduced into a set or removed from it; but there is no way, in the formalism, for new elements to be created, nor for any to be annihilated, in any elemental interaction or other possible occurrence. (This is to be justified below.)

Each element name contains, is constituted by, the *history* of all of the element's previous interactions, without a break, starting from a certain moment in its history, taken as significant for a particular discussion, through its most recent interaction. The elements partaking of an interaction, thereby entering into each other's histories, will be said to *determine* each other to the extent of that interaction. Each elemental interaction involves two determinations, one for each of the elements engaged. The interactions represented, *when related to the history of preceding interactions,* determine the element's *state* at the instant of the most recent interaction. Each interaction is a stage in the progressive determination of states in an element's history. As is evident, an element, its state at a given moment, its interaction with a particular other element at that moment, the history of that and of previous interactions, are all implicated in each other; they are all represented by the element name, by the precise way it is formulated; they are different ways of viewing and understanding the element's name.

An element's name, that is, the representation of its known, or significant, history, doesn't correspond to anything *in* an element, since there can be nothing "in" it, being, as it is postulated to be, utterly simple, uncomposed, without parts or aspects, without interior or exterior, without any other property, to which the element name, in all its (potentially limitless) complexity can be attached. Its history is not contained in, or imprinted upon, an element. Differences in element histories (therefore names) have significance as indicating the size and character of the set of elements of which particular elements are members. The character of a set itself boils down to the number of other elements with which any given element has interacted or can interact in a particular sequence of interactions, and the probabilities associated with each of the possible interactions.

Though all elements are utterly simple, they can have the most complicated histories, and therefore a corresponding diversity of states. Among the most basic distinction of states — and I must, because of the purpose of this work and my own limitations, confine myself to the most basic — are those between *reciprocal* states (e.g., /0/00//00/0//) and *non-reciprocal* states (e.g., /0/00//000/0000//); between *symmetric* states in which the immediately preceding states of both interacting elements had been either reciprocal or non-reciprocal, and *a-symmetric* states, in which the immediately preceding state of one element had been reciprocal while that of the other had been non-reciprocal; among *occurrent* states, in which new elements appear in a given element's history, *recurrent* states, in which only elements that had previously occurred keep recurring, and *mixed* states, in which both occurrences and recurrences appear in an element's history; among mixed states, *periodic* states, in which occurrences and recurrences are at regular intervals in the element's name and *a-periodic* states, in which the intervals are irregular. Runs of occurrent determinations can be of varying lengths. The number of such determinations in unbroken series in the immediate past history of an element establishes the *level of development* of its latest state. A recurrence arrests development. A reciprocal interaction causes *degeneration* of state to the *null-level,* and the reciprocating dyad will continue at the level until a non-reciprocal interaction occurs. So, too, a-symmetries can be of varying degree, depending on the respective levels of the two elements involved in an interaction; and periodicities and a-periodicities can vary still more greatly.

The properties of sets of elements depend on the properties of the elements and interactions that constitute particular sets. Elements that are recurrently available to each other (that recur in each other's histories) form a *cluster.* Clusters in which occurrences and recurrences are periodic constitute a *group.* Clusters and groups in which the ratio of occurrences to recurrences increases *develop* during the period of such increase. Conversely, clusters and groups in which the ratio of occurrences to recurrences decreases *degenerate* during the period of such decrease. Where there is no recurrence, only occurrences, the cluster or group *disintegrates.* Elements only occasionally and irregularly available and those not themselves available to given elements (not *directly* entering their histories), but *indirectly* available to given elements, as having entered the histories of elements with which the given elements interact, constitute the given element's *field.* The *periphery* of a cluster or group comprises those elements relatively most involved in field interactions. The *center* of a cluster or group comprises the elements less involved in field interactions. (Obviously, centrality and peripherality are matters of degree and can change over time.) A cluster or group in which center and periphery show increasing interactions tends toward *integration* and, at the extreme, *coalescence*. To the extent that the periphery's field interactions tend to preponderate over interactions with the center, the cluster or group tends to *disintegrate*. The number of possible kinds of groups, clusters, and fields is a very large multiple of the kinds of individual element states. (Some of these kinds, e.g., strings, loops, kinks, chains, knots, braids, webs, and their transformation through time, will be touched upon in the next chapter.) As the number of elements in a given group, cluster, or field increases, the possible variations in their patterns of interaction increases exponentially and, accordingly, so does the possible variety of forms and transforms of the sets. The set of interactions that tends to change the properties of a group, cluster, or field is a *process*, which can itself vary enormously in rapidity, in degree of pervasiveness, and in many other ways.

The entire system of valid element and elemental interaction representations constitutes an *object language*, the most fundamental of object languages, since the objects represented are the postulated ultimate elements and interactions; the only one, so far as I know, in which all the terms have histories, in which the terms *are* histories.*

*Physicists first recognized the importance of time — that is, its a-symmetry, "the arrow of time," which is its most fundamental property — with the formulation of the second law of thermodynamics. In Relativity Theory, time, as the fourth dimension, is intimately involved in all physical processes, is dependent on the observer's reference frame, and is therefore variable. In quantum theory, not only time, but history, that is, the record of passing time, has great importance. "In classical physics, the history of a particle is essentially irrelevant to the understanding of the behavior of the particle. . . . In quantum theory, this is not so. . . . The quantum state always refers back to earlier states, and can be regarded as a specification of the information we have about the particle" (T. Bastin in *Quantum Theory and Beyond*, edited by T. Bastin, Cambridge Uni-

The object language presented *exhibits* (rather than merely referring to) the possible forms and transforms of element states and element sets (groups, clusters, fields), accordingly groups, clusters, and fields of histories. In showing, rather than just asserting, it resembles the formulas of stereochemistry, the mappings of the earth and the heavens, electronic circuit designs, computer programs, flow charts, structural blueprints, musical scores, and the like, rather than the equations of mathematics, the propositions of formal logic, or the sentences of ordinary speech. It contains no equations or propositions. It makes no assertions.*

---

versity Press, 1971). In R. Feynman's version of quantum electrodynamics, the behavior of an electron depends on the summation of all its possible histories, "the sum-over-histories," not only what (is presumed to have) actually happened to it, but what might have happened, as constrained by such features of quantum mechanics as the cancellation of possibilities, so that the probability of a certain kind of behavior may diminish, rather than increase, as the number of possible ways it could happen increases. However, in Quantum Theory (whether Feynman's version or those of others) as also in Relativity Theory and in statistical mechanics, histories are only proposed or adumbrated. Nowhere is there any attempt to spell out the history, moment by moment, for all the elements of a particle. The object language here presented attempts to represent histories fully and rigorously — though, for reasons already stated, such a program can in practice only be realized for very few elements over very short sequences of interactions.

*For this reason, the formalism avoids the paradoxes that plague mathematical logic and that the Theory of Types and the Axiom of Reducibility do not entirely remove. Nor are the demonstrations (first by K. Gödel, and then, in different terms and contexts, by Turing, Church, Tarski, Post, and others) that it is impossible to prove the completeness and logical consistency of a system of sufficient complexity to make it interesting, relevant to my formalism. Such demonstrations apply only to mathematical and logical systems in which statements can be made, particularly reflexive ones, about the properties of a system or of constituent aspects of its. A meta-language, whether artificial or natural, used to describe representations of elements and their interactions, and making statements about them, would, however, be subject to the difficulties exposed by Gödel and the others.

Like the "picture language" that Wittgenstein sought, but could neither find nor construct, the formalism here presented *shows* possible element states and interactions and their possible grouping, clustering, and field envelopment. It makes them manifest and, as such, *occasions* for possible assertions, whether true or false, about them. Such assertions would be in a meta-language, either some natural language or some mathematical formalism.

Exclusions

While the choice of the characters used in the formalism and its other non-structural characteristics (e.g., adding a stroke at the end rather than at the beginning of a merged expression) are arbitrary, the claim is made that the construction itself, its truly structural features, is a uniquely necessary consequence of the Rule of Unambiguous Representation and its sub-rules. For inferences for objective reality to be validly made from the formalism, its construction *must* be just so and not, in any respect, however apparently trifling, otherwise. This has to be rigorously demonstrated. The only way I know to do so is to exclude all possible alternative constructions, as inconsistent with the rules.

There are two kinds of difficulties here. The first is that I may have overlooked certain possible structural variants — and accordingly failed to invalidate them — or that the proofs offered are not perfectly rigorous. It is for the discerning reader to judge whether there are significant omissions or insufficient rigor. The second difficulty is that the rest of this chapter is inescapably tedious, especially since, for the sake of the utmost rigor of which I am capable, I present the exclusions in the stilted and redundant terms customary in formal proofs. The reader is, of course, free to skip the rest of this chapter and, provisionally taking my word for it that the eliminations are sufficiently complete and the proofs sufficiently rigorous, go on to see what inferences would follow from the formalism and what, if any, interest they may have. With these caveats, herewith the exclusions:

Given:

The Postulate of Elementality: "Matter," "energy," "space," "time," the "fundamental forces," whatever there is that we can know of physical reality, are all composed of elements, the ultimate constituents of all physical structures, and of elemental events, the ultimate constituents of the processes through which structures change — elementality being defined as utterly simple, therefore uncomposed, unarticulated, indivisible, irreducible, devoid of magnitude or any other property except simplicity.

The Corollary of Identity: Elements are intrinsically identical with each other; elemental events are intrinsically identical with each other. However, elements differ from each other extrinsically, that is, with respect to the histories of the elemental events in which they had participated, and in no other way.

The Corollary of Action: All elemental events are actions of particular elements upon particular other elements. There is no other kind of elemental event.

The Corollary of Availability: Every element has at least one other element, but not all other elements, available for interaction with it.

The Corollary of Dyadal Interaction: Any element can act upon only one other element at any one time, and must so act; and the element acted upon can, and must, at the same instant, react upon the acting element, and no other; so that elemental events are all dyadal interactions.

The Principle of Otherness: Given any two elements, available to each other, each is not the other, even though they are intrinsically identical. Given any two elemental events, each is not the other, even though they are intrinsically identical.

The Rule of Unambiguous Representation: Since elements and elemental events are just what they are and not other than they are, they must be represented unambiguously, each element or event in only one way, and each other element or event in distinct other ways, each other than the others, so that it is never, in principle, possible to mistake a particular element or event for any other, to confuse one with any other, to take or understand any of them as other than they are. As for individual elements and events, so also for sets of elements and events formed by the combination and permutation of elements and events. Each set must be unambiguously represented, so that it is never, in principle, possible to mistake one for any other, confuse one with another, to take or understand any of them as other than they are.

The Sub-Rule of Dyadal Representation: Each elemental event represented must involve the interaction of two elements, never any more or any fewer.

The Sub-Rule of Representing Histories: Elements differ from each other only extrinsically, that is, in the histories of their interactions with particular other elements. The extrinsic non-identity of particular elements, accordingly their respective histories of interactions with particular other elements, must be represented unambiguously, therefore uniquely (since variant representations could be variously understood).

The Sub-Rule of Unique Representation: No expression in the formalism may represent more than one element state; conversely, no element states may be represented by two or more expressions, but each only by one.

The Constraint of Sufficient Reason: How we represent particular elements and their interactions, and the combinations and permutations of such elements and interactions, and what we may legitimately say about any of them must be limited to what can be validly inferred from the uniquely necessary formulation of what is implied by the Postulate of Elementality and its corollaries, the Principle of Otherness, and the Rule of Unambiguous Representation and its sub-rules. Nothing may be assumed to exist or to occur and to become any more or any less probable without sufficient grounding in the formalism and the premises on which the formalism rests.

From these premises there derives:

Fundamental Theorem: There exists a construction uniquely capable of representing all elements and all elemental events, and all possible combinations and permutations of such elements and events. This construction must take the rudimentary form

/0/00/   /00/0/

and its elaboration, as indicated in the preceding section — the elaboration depending on the number of elements and elemental interactions, the order of interactions, and the possible combinations and permutations of elements and interactions — and no other form.

1. *Excluded*:   The use of only one character (or other signal) rather than two or more, to represent elements and elemental events.

The proof: Suppose we represent the successive interactions of a pair of elements thus:

0 00   00 0
0 00   00 0

In the second set of expressions above (that is, those on the second of the two lines of type) the second interaction and the resulting states of the two elements is represented in the same way as the first, which is contrary to the Sub-Rule of Unique Representation, which stipulates that no expression in the formalism may represent more than one element state. The second interaction (two reciprocal actions) and the resulting states, as represented above, are identical with the representation of the first interaction and the resulting states, which is contrary to the Rule, therefore excluded.

Suppose now that, to obviate this objection, we represent two successive interactions of an element couple thus:

$$0 \qquad 00$$
$$00 \qquad 0000$$

This, however, violates the Sub-Rule of Representing Histories, which requires that the history of preceding interactions be unambiguously, therefore uniquely, represented. The second set of expressions does not unambiguously represent the history of the preceding interaction of the pair of elements involved.

Suppose that, to meet this difficulty, we write

$$0 \qquad 00$$
$$000 \qquad 000$$

to indicate that, in the second interaction, element 0 had previously acted on element 00, and that 00 had previously acted on 0. But then the two element states resulting from the second interaction are represented in the same way, which is contrary to the Sub-Rule of Unique Representation, which interdicts representing extrinsically different element states in the same way.

Actually, this alternative, the use of only one character rather than two or more, is specious, since the space separating the characters used, the 0s, is itself a signal, no less than the / proposed for my formalism. Absent the space, there would only be an unbroken succession of 0s, obviously quite meaningless. (As previously noted, with one character nothing can be said; with two, everything.)

2. *Excluded*:   The use of three or more characters (or other signals), instead of only two, to represent a series of

interactions — accordingly 0/000 for the first action of 0 upon 00, then, say 0'00', or $0^2/00^2$ or $(0/00)^2$, or whatever, for the second action of 0 upon 00.

The proof: As already indicated in introducing the formalism, if three characters are employed when two would suffice, there would be more than one way of representing the same element or the same event, the two-character way, a variety of three-character ways, a still greater variety of four-character ways, and so on. But this is contrary to the Sub-Rule of Unique Representation.*

3. *Excluded*: Representation of elements singly.

The proof: Excluded by the Corollaries of Action and dyadal interaction. All elemental events are actions of particular elements upon particular other elements. Nothing can happen to an element internally; it has no "inside." There can be no self-action of an element, for that would imply that one of its parts or aspects acts on other parts or aspects. But an element, being utterly simple, therefore uncomposed and indivisible, has no parts or aspects. Nor can time pass for an element without participation in interactions with other elements. An element represented singly could not, accordingly, represent any possible event.

---

*But what, it may be asked, about the space between characters and between lines of type as a third kind of signal. They would indeed be a third kind of signal, and therefore contrary to the Sub-Rule of Unique Representation, were they structrual features of the formalism. They *could* be given a structural role, substituting for the /s so that there would be two spaces, or three, or whatever, between characters instead of two or three or more /s. But I have arbitrarily chosen to employ /s as being a more distinct kind of signal. As it is, the spaces separating expressions on a line and those separating lines of type have no structural significance; they can accordingly be as large or small as the author or the typesetter chooses. They function like scissors used to cut a frame from a reel of movie film, creating a "still," a frozen moment in a continuous sequence of events.

4. *Excluded*:   Representing three or more elements as engaged in one mutual interaction.

The proof: Suppose an interaction is represented thus:

00	0	0
0	00	000
000	000	00

But that would imply that an element, say 0, has parts or aspects, one acting upon element 00, the other upon element 000; and similarly for the other elements involved. But that is contrary to the Postulate of Elementality and its Corollary of Dyadal Interaction. Also, there would unavoidably be a number of different possible ways of representing an interaction involving three elements (and the more so for those involving more than three). For example, instead of the way shown above, the following way

000	000	00
0	00	000
00	0	0

But that, and the other possible variants, would be contrary to the Sub-Rule of Unique Representation.

5. *Excluded*:   Non-accretion of expressions.

The proof: Without such accretion, the identity of the elements participating in a series of interactions, and therefore the precise history of interactions — indicative of just what other elements a given element had interacted with at a certain moment — would be left ambiguous, which is contrary to the Sub-Rule of Representing Histories. Suppose element 0 acting first on 00 and then on 000. Were we to represent the first action as 0/00 and the second as 0/000, it would not be clear, from the second expression, that element 0, most recently acting on 000, had, at the immediately preceding moment, acted upon 00, rather than upon 0000 or 00000 or any other element. Elements differ only in their histories; different histories mean different extrinsic identities; and different identities must, by the Sub-Rule of Unique Representation, be differently represented, and each particular identity uniquely represented. Therefore, if element 0 acted first on 00 and then on 000, that must be shown as 0/00/000 — or, properly, for reasons not yet fully established, as /0/00//000/0000// — which may be read: element 0, which *had* acted on element 00 *and no other*, is *now* acting on element 000.

6. *Excluded*: Presenting strokes singly, as /0/00/000/ 0000/, for example.

The proof: The expression shown (and all others like it) is inescapably ambiguous, which violates the Rule of Unambiguous Representation. The expression could be understood as "element 0 acts upon element 00, which *had* previously acted on element 000 and previous to that on element 0000" or, alternatively, "element 0, which *had* acted on element 00 and previous to that on element 000, is *now* acting on element 0000" and in several other possible ways, the number of possibilities rising exponentially with increasing lengths of element names. But that would mean obscuring the precise sequence of actions, thereby the history and the resulting state of element 0, which is contrary to the Sub-Rule of Representing Histories; and it would also mean that the identity of element 0, which depends on its history, would be variously represented, which is contrary to the Sub-Rule of Unique Representation. Only the bunching of /s — the number of /s in each bunch corresponding to the order of successive actions — can make the sequence of events, therefore the histories and the resulting states, clear.

7. *Excluded*: Adding or interpolating varying numbers of /s (the number depending on the order of the actions represented) at appropriate positions in element names.

The proof: There is only one kind of elemental event, the action of a particular element upon one (and only one) other element, and the reciprocal action of the second element upon the first — both together constituting an interaction, of which each is an aspect, one from the perspective of the first element, the other from the perspective of the second element. (All other events and processes in nature are assumed, by the Postulate of Elementality, to be composed of such elemental actions.) All elemental actions are, by the Corollary of Identity, intrinsically the same. While elements have histories of actions and are accordingly extrinsically different, elemental events have no histories — instead they constitute histories — and therefore they cannot differ extrinsically. It is always the same action. Each action is given its unique character not by the differences in the action abstractly considered — there can be no such differences, by the Corollary of Identity — but only by the histories, therefore the different extrinsic identities of the elements involved in it. Since elemental actions, considered apart from the elements participating, are all and always the same, it is impermissible, by the Sub-Rule of Unique Representation, to represent them in different ways. Only a single / (or whatever other signal may have been arbitrarily chosen) may be used to represent an elemental action considered abstractly, that is, when not yet incorporated in element names — which, through such incorporation, occur not only singly but also in bunches of varying length depending on the order in a given sequence of actions.

8. *Excluded*:   Placing the added stroke between the names of interacting elements when juxtaposing the names, instead of at the end (or the beginning) of the resulting expression.

The proof: Suppose the first interaction of elements provisionally named 0 and 00 were represented as

0/00   00/0

and the second interaction between the same elements as

0/00/00/0   00/0/0/00

As is immediately evident, there would then be no way of unambiguously understanding the histories of the interacting elements, the precise order of interactions. But this violates the Sub-Rule of Representing Histories.

Suppose now that we represent the first interaction as

/0/0/   /00/0/

and, interpolating single strokes between the juxtaposed names, the second interaction as

/0/00///00/0/   /00/0///0/00/

But, since this would mean that there would be three /s in the middle of the names resulting from the second interaction, it would render the histories of the given elements and the resulting states ambiguous, contrary to the Rule of Unambiguous Representation. This becomes even clearer from representing a third interaction between the same pair of elements, again interpolating a single stroke between the names of the elements,

/000///00/0///00/0///0/00/   /00/0///0/00///0/00///00/0/

in which the third interaction is represented by the same bunching of strokes in threes, as the second, resulting in the confusion prohibited by our construction rules.

Suppose again that we represent the first interaction as

/0/00   /00/0

and, again interpolating single strokes between the juxtaposed names, the second interaction as

/0/00//00/0   /00/0//0/00

Continuing the procedure, the third interaction would be represented as

/0/00//00/0//00/0//0/00   /00/0//0/00//0/00/0

As is clear, the third interaction comes to be represented by the same two strokes as the second, which is against the Rule of Unambiguous Representation.

Whichever way one goes — and I have tried out all other possible variants I could think of — one ends up with ambiguous representations of element histories, which violates the Rule of Unambiguous Representation and the Sub-Rules of Representing Histories and of Unique Representation. The only way of satisfying these rules is the way adopted, adding single strokes to the end (or the beginning) of juxtaposed element names.

9. *Excluded*:   Histories, accordingly names, of interacting elements of different lengths.

The proof: Were names of two interacting elements of unequal length, it would mean either that the interacting elements had come into being at different times (and are accordingly at different "ages") or that, during the same period, the two elements had been involved in different number of actions.

The second possibility is excluded by the Corollary of Action, which excludes all elemental events except the action of particular elements upon particular other elements, and by the Corollary of Dyadal Interaction, which requires that every element act upon another element, and only one other, at any one time, while the element acted upon must, at the same instant, react upon the first element, and no other. As explained in Chapter 2, there is no time lapse, no time, without elemental events, accordingly elemental interactions. No time can pass for an element without interactions. And since all elemental actions are without duration, instantaneous, for a given number of instants, there must be equal numbers of elemental actions. While, as we shall see in the following chapter, the time passed *at different levels of development* of mass–energy will depend on the rate of motion of particular aggregations of elements relative to other aggregations,* apart from consideration of level of development, time must pass uniformly for all element in the universe, everywhere in the universe.

The first possibility, that elements might be of different ages, is excluded by the Constraint of Sufficient Reason, which prohibits representing elements and their interactions, or saying anything about elements and their interactions, except what is unambiguously inferable from the premises of the formalism. Nor may anything be asserted of different sets of elements or interactions, except as inferable from permissible combinations and permutations of elements and interactions. We cannot derive from the Postulate of Elementality and its corollaries, the Principle of Otherness, and the Rule of Unambiguous Representation and its sub-rules any inference about the creation or annihilation of elements. Accordingly, we may not represent, or imply by representation, the creation or annihilation of elements. All we can represent are minute segments of element histories, which, for all we know, stretch backwards and forwards eternally. The universe, as a particular kind, or set of kinds, of *organization* of elements and elemental events, may have had a beginning and may have an end; but its constituent elements are, from the premises of this formalism, eternal.

Because there is no warrant for representing elements as being of different "ages," and none for holding that different number of interactions may occur in the same time frame, the elements in an interaction must be represented as being of the same age and as having undergone the same number of interactions in the past. And the elements must accordingly be represented with histories, therefore names, of the same length.

* * *

*This is the basis, as we shall see, of the "time dilation" of Relativity Theory.

So much for the demonstration — incomplete and imperfect though it may well be — of the formalism presented above. Establishing its unique necessity would contribute Specificity to the conditions of intelligibility. Granted the premises from which the formalism derives: The Postulate of Elementality, which is a reasonable conjecture about the fundamental nature of objective reality — more reasonable, I submit, than any of its alternatives; the Principle of Otherness, which, being analytical, is incontrovertible, and the rules for enforcing that Principle, the formalism just presented follows uniquely and necessarily, and is accordingly isomorphic with the basic structures and processes of nature.

Still to be satisfied are the conditions of Connectivity and Adequacy — and the Beauty that would result were all the conditions for intelligibility met. That is the task for the next, the final, chapter.

Chapter 4
Models and Explanations

The Postulate of Elementality posits that elements and elemental events (as therein defined) are the ultimate constituents of physical reality; that all natural structures and processes are combinations and permutations of elements and elemental events. It would follow that any formalism that can be shown to derive, uniquely and of necessity, from this postulate (together with its corollaries, the Principle of Otherness, and the Rule of Unambiguous Representation and its sub-rules) would — if the premises are accepted and the formalism correctly derived — serve to mirror all structures and processes in nature. The means of such service would be the building of models of particular combinations and permutations of its notation. Such models would be isomorphic with the structures and processes modeled. If isomorphism can be established, inferences about natural phenomena could be drawn from such models, as inferences about a terrain can be validly drawn from a map known to be isomorphic with it, or inferences about the properties of the DNA molecule, from an accurate three- (better four-) dimensional model. My endeavor should, accordingly, be to construct models employing my formalism that precisely reflect selected natural phenomena, and so to make the phenomena intelligible in terms of the premises here advanced.

However, in attempting to construct such models, I am faced with a number of serious difficulties.

First, the premises of the system have to be accepted — and they need not be. (The claim of incontrovertibility is made only for the Principle of Otherness.) Also to be accepted is that I have completely and rigorously established the uniquely necessary derivation of my formalism from my premises. (Given my personal limitations, I can have no perfect assurance of completeness and perfect rigor.)

Second, what has to be modeled in terms purported to represent "objective reality" is not that reality, but phenomena, that is, the appearances of phenomena under observation, experiment, and measurement by human beings whose powers of perceptions are severely constrained, who are notoriously prone to fallibility within those constraints, and who use instruments of observation, experiment, and measurement that are unavoidably imprecise and inconsistent (that is, they give different readings at different times) and that are often improperly designed and improperly used. More important is the consideration that the conceptions of what is being observed, experimented with, and measured tend to be uncertain and mutable. Confirmed "facts" tend to get mixed up with hypotheses, theories, conjectures, speculations, paradigms, the patterns of thought inherent in language, all languages including those of mathematics. There are no naked facts, as W. V. Quine showed in *From a Logical Point of View* (2nd edition, Harvard University Press, 1961). Thus, notions about "elementary" particles, their interactions, and the fundamental forces that are presumed to be embedded in the particles or to govern their interactions have become, in successive versions of quantum (and related) theories, less and less clear, less and less accessible to common understanding, more and more problematical. What is worse, the conceptions keep changing, so that particles a short while ago thought of as "elementary" are no longer so regarded; elementary particles until very recently universally conceived as points are now, by an increasing number of theorists, analyzed as "superstrings"; interactions once thought of in terms of classical fields (the way Faraday and Maxwell conceived them) are now said to depend, in all instances, on interchange of particles. I could go on. Therefore, a model that I might propose to fit some current conception of a particle, an interaction, a force, or other aspect of physical phenomena, may prove irrelevant to some revised view or views of what is to be modeled that may emerge before this work is read.

Exacerbating the difficulty of modeling phenomena is the consideration that physics and mathematics (the preferred, indeed the indispensable, language of physics) are quite different disciplines, with different traditions, concepts, procedures, standards of rigor, etc., so that the same terms can have different meanings or, what is worse, the relationship of the meanings intended by mathematicians and physicists, respectively (or by sub-groups in either discipline) is obscure. For a pure mathematician, a term's meaning depends entirely on its relation to other terms in a particular abstract system, and on nothing else. For a physicist, a term's meaning also depends on its relation to observed data — in their concreteness, fundamentally different from mathematical abstractions. As M. Eigen and R. Winkler warned in *The Rules of the Game* (New York: Alfred A. Knopf, 1981): "Abstractions like point, infinity, continuity, parallelism and so on are standard conceptions in mathematics but in physics they have to be applied with great caution."

A third set of difficulties lies with the proposed formalism itself. As already indicated, it is exceptionally — indeed, by its very nature, uniquely — cumbersome and unwieldy. Since an elemental interaction occupies no more, and maybe a good deal less, than 10^{-43} seconds, I have tentatively calculated that modeling an electron over the course of a second — which would mean spelling out in full the names (that is, the relevant histories) of all the elements constituting an electron and of all the field elements with which they interact, for all the instants, that is, all the interactions, of a second — would require on the order of a quintillion times all the volumes of all the world's libraries. It follows that literal modeling, in the sense of strict isomorphic representations of the phenomena modeled, is just not feasible. I must therefore describe, rather than actually represent, the elements and the interactions involved in the phenomena to be modeled, using for the purpose some meta-language to produce what might be called "meta-models."

This leads to the fourth set of difficulties, those concerning meta-languages. The meta-language employed here is ordinary semi-scholarly English. But, of course, any natural language, being the resultant of a vast array of diverse, often conflicting, influences is, by its nature, imprecise, vague, ambiguous, mutable, despite all efforts at remedy by grammarians, lexicographers, and other purists and reformers of language. They are, therefore, ill-suited to the clarity and rigor required for scientific, and other scholarly discourse.* Many artificial verbal languages have been proposed, which would be less ambiguous, more steady, therefore more reliable, than any of the natural languages. None has won significant acceptance.

---

*Pertinent here is the comment of G. Steiner in *Language and Silence* (New York: Atheneum, 1967): "It is arrogant, if not irresponsible, to invoke such basic notions in our present model of the universe as quanta, the indeterminacy principle, the relativity constant or the lack of parity in so-called weak interactions of atomic particles, if one cannot do so in the language appropriate to them, that is, in mathematical terms. Without it, such words are phantasms to deck out the pretences of philosophers or journalists. Because physics has had to borrow them from the vulgate, some of these words seem to retain a generalized meaning; they give a semblance of metaphor. But this is an illusion." He goes on to argue that one cannot "meaningfully *speak*" [his emphasis] of "the operations of Lie groups or the properties of n-dimensional manifolds or the complexities of topological analysis."

The language of choice among physicists for all the most significant portions of their papers (that is, those most precise and unambiguous, and those that most readily lend themselves to testing and to rigorous logical and mathematical manipulation) is, of course, mathematics. Very much more satisfactory as meta-languages describing and discussing my models would be some branch or branches of mathematics (which themselves, incidentally, are object-languages for a variety of meta-languages, e.g., the meta-mathematics formulated by G. Frege, B. Russell, D. Hilbert, and those stemming from them). I have found various branches of algebraic topology (particularly lattice theory and knot theory) especially promising. But here again, there arise two kinds of difficulties — the first having to do with the limitations of topological analysis, the second with my own limitations.

As to the first; Physicists who have tried to apply topological analysis to physical phenomena have been frustrated not only by the present relatively undeveloped state of topology but by its very nature. For,

> Topology cannot actually solve equations. What it provides is a mathematical vocabulary — adjectives and nouns — that allow a set of solutions to be discussed in a general way without actually being specified. Thus, although the manifold of points that make up the set of solutions to an equation has a precise and unambiguous shape, the topology of that manifold are not constrained by the properties of that shape. Instead the topology encompasses whatever properties are retained when the manifold is deformed in an arbitrary way, as long as the deformation is done without cutting, tearing or puncturing.

(W. P. Thurston and J. R. Weeks, *Scientific American,* July, 1984). Comment by Nobel Laureate A. Salam (in *Scientific Explanation, op. cit.*) is also pertinent:

Wheeler, Schemberg and Hawking associate the electric and
nucleonic charges to space–time topology — space–time
Gruyere-cheeseness, wormholes of the granular size of
10^{-33} cm. The idea is attractive. Topology, you may recall, is
concerned with "global" aspects as contrasted with the "differential" aspects of the present tradition of physics. It thus
represents a real break with the past. Unfortunately — and
I say this deliberately and ungratefully in order to provoke
— my own feeling is that the thematics of topology in respect
of what we need has not progressed beyond the Mobius strip
and the Klein bottle. Topology, as a language for physics, is
not yet capable of supporting the edifice the physicist may
want to erect on it. Could it be that our generation will be
defeated by the lack of development of a necessary mathematical discipline in a direction that we need.

Because topology is non-metrical, therefore non-quantitative, a topological meta-language cannot in itself provide
theories that might be fully satisfactory to "the present tradition in physics." "What it provides is a mathematical vo-
cabulary — adjectives and nouns — that allow a set of so-
lutions to be discussed in a general way without being
specified." That could be worth a good deal, were the quality of the non-quantitative discussion sufficiently good. It is
not yet so. The trouble with the "wormholes," the ten-or-
more-dimensional manifolds (of which six or more dimensions are invisible), like the "hidden-variable" theories I
have earlier discussed, is that the concepts employed and
the models generated are ad hoc contrivances. They are designed to fit, "forced to fit" the known phenomena, but they
are not themselves grounded in basic principles of natural
order. "All science — physics in particular — is concerned
with discovering *why* things happen as they do. The 'why's'
so adduced must clearly be 'deeper,' more universal, more
axiomatic, less susceptible to direct experimental testing,
than the immediate phenomena we seek to explain" (A.
Salam, *op. cit.*). His feeling (which I share) is that the topological models so far advanced lack the qualities he lists.

It is my claim that the models soon to be presented are "deeper," "more universal," "more axiomatic," because of the nature of the premises and the formalism on which they rest, than those advanced by those physicists who have applied topology in formulating their theories and building their models, even though as with other topological models, they are not "susceptible to direct experimental testing." They would afford a conceptual framework, rather than a theory rivalling the currently accepted ones, through which such theories might be made intelligible. The particular models to be presented are hypotheses to be tested against current theories and the phenomena they seek to organize (and, in that limited sense, to explain), in order to determine whether the patterns exhibited in my models and the patterns thought to characterize the phenomena being modeled do indeed correspond.*

There remains the final, and perhaps most crucial, difficulty: my own limitations. I am not enough of a topologist (or any other kind of mathematician) to be able to construct fully and rigorously worked-out topological models (or, for that matter, those that might be couched in terms of some other branch of mathematics). What follows are but hints and gropings, clumsy and unfinished, for models. I make use of certain topological concepts, but I express them of necessity in ordinary language, aware though I be of its imperfections. The best that I can do is to make a start and then to point the way for those more competent to investigate. It is for my readers to decide whether or not that way is worth pursuing.

* * *

---

*Ideally, my models and explanations should be compared only to confirmed "facts" and not to the hypotheses, theories, speculations by which "facts" are ordered. Unfortunately, as I have already noted, facts cannot be cleanly unscrambled from their conceptual matrix.

Basic Space–Time Models

I begin with a proposed model of a *space–time point*: one
reciprocal interaction of one elemental dyad.*

A succession of reciprocal interactions of the elements
of the same dyad would constitute a *point-duration,* the
length of the duration depending on the number of such
interactions in unbroken sequence. Since, by the Corollary
of Identity, all interactions are intrinsically identical with
each other (however they may differ extrinsically); and
since, by the corollary of Dyadal Interaction, every element
must interact with some other element at every instant, with
the consequences of unbroken series of intrinsically identical interactions, durations involving the same number of interactions are all equal. (This is shown in the formalism, in
which interacting elements all have the same number of
strokes in their names.) If the interactions are reciprocal in
unbroken sequence, all such sequences of equal number oc-
cupy equal point-durations.

It may be objected: How can durations, which have
magnitude, be formed out of interactions that, by definition
of my Postulate of Elementality, have none, being instantaneous. The sum of however many zeros would still be zero.
However, while it is true, by my Postulate, that an interaction has no magnitude, it may not be thought of as a zero,
but as a one: each one being other than any other one, accordingly two ones, then three, four, etc. One instant fol-

---

*It is tempting to equate my space–time point with Planck's unit of ab-
solute space–time, to which the value 5.36×10^{-44} seconds is usually as-
signed. But Planck arrived at this unit by relating his constant $\hbar$ to that
of the velocity of light c and Newton's gravitational constant G. It is , to
my mind, a speculative and rather arbitrary derivation. Furthermore, it
has mischievous connotations, as suggesting divisibility into still smaller
magnitudes, e.g., 2.68×10^{-44} or 5.36×10^{-45}. In any case, the relationship, if any, between the Planck unit of space–time and my own has no
direct bearing on our discussion.

lowed by another is more than one instant, therefore a quantity of time, a duration.

A sequence of interactions, of whatever magnitude, constitutes a *continuum,* in the sense that nothing does, or can, intervene between interactions except other interactions; but not in the classic mathematical sense of infinite divisibility — in which there would be an infinity of interactions, therefore instants, between any two given interactions, however close in time, even closer than Planck's hypothetical time-unit. The mathematician's continuum has — as is now recognized by physicists, with their virtually unanimous acceptance of the quantized view of reality — no counterpart in nature. (The rejection of this fiction enables us, as we shall see below, to resolve certain of the paradoxes of contemporary physics, as well as old Zeno's paradoxes and those arising from G. Cantor's Chinese boxes of infinities.)

It might appear that space–time consists of the totality of space–time points. But this would be a hasty, and mistaken, extrapolation. For reasons to be discussed below, we can speak of *all* of space–time, or *universal* space–time, or of the universe *as a whole,* as distinguished from local times and spaces of which we can have direct knowledge, only by highly tenuous inferences from a complex of theories and theory-laden "facts" — some of which can be justified in terms of my models, while other cannot.

A point-duration has no relationship whatsoever to any other point-duration, so long as it remains a point-duration, that is, an unbroken succession, of whatever length, of *reciprocal* interactions. It cannot legitimately be said to be earlier or later or contemporaneous with any other point-duration. Nor, by the same token, can a point at any one instant be related spatially to any other point. It cannot be said to be next to another point, to be near to or far from it, or to have any other specifiable relation to it. It is only when the constituent elements of a point interact *non-*reciprocally, that is, with elements other than their partners, thereby *annihilating* the point (the dyad when reciprocating) they composed, that relationships can occur (and be validly predicated) between the elements of a given dyad to the elements with which they come to interact non-reciprocally, with the elements that occur in the histories of the non-reciprocal elements involved, and with other elements inferable from such histories as being, or as likely to become, available for interaction with the elements of the given dyad.

I submit that space is continuously being formed and reformed by the creation, annihilation, and re-creation of points. Since points do not relate to each other except through their annihilation in non-reciprocal interactions, there is no space except by inference from the *histories* of points. (Such inferences may be purely theoretical or they may be intuitively compelling, as when we "experience" distance because of objects intervening between ourselves and the object of our special regard, because of diminution of size, blurring of outlines, reduction of brightness, alterations in color, the effects of parallax, and numerous other "clues" to distance, of which master artists have made exemplary use in creating the illusion of a three-dimensional space on a two-dimensional surface.) The elements that constituted a point that has been annihilated by non-reciprocal interactions do not again become constituents of a point until, in a series of interactions, reciprocal interactions again occur. There is no space except where there are reciprocal interactions following a series of non-reciprocal interactions. Again: there is no space except through the annihilation and re-creation of points.

The Dimensions of Space–Time

At any given moment, a space–time point, that is, a reciprocating dyad, has five possible kinds of interactions available to its constituent elements.

1. Another reciprocal interaction, involving the same pair of elements. This perpetuates the point for another instant. It corresponds to the *null dimension*.

2. A switched occurrence, in which one element of a reference dyad interacts with one of the elements of an occurrent dyad (that is, a dyad the elements of which had not previously entered the reference dyad's history), while the other element of the reference dyad interacts with some other occurrent dyad. The two occurrent dyads are assumed not to be available to each other, because the reference dyad intervenes, lies between them, thus:'

Reference dyad

(3) /000/0000/ ⟷ (1) /0/00/   (5) /00000/000000/ ⟷

⟷ (4) /0000/000/   (2) /00/0/ ⟷ (6) /000000/00000/*

Since the two occurrent elements are assumed not to be available to each other, the unengaged elements, labeled (4) and (5), respectively, cannot interact with each other; but must, at the very moment their partners are engaged, interact with members of other occurrent dyads; and so on for a string of interactions radiating out in opposite directions from the reference split pair. This corresponds to the *first dimension*, that of the line. The significance of switched occurrent interactions will be elaborated in the electromagnetic radiation model below.

3. A paired occurrence, in which one element of a reference dyad interacts with one of the elements of an occurrent dyad, while the other element of the reference dyad interacts with the other element of the occurrent dyad, thus:

/0/00/ ⟷ /000/0000/
/00/0/ ⟷ /0000/000/

Instead of a line of dyads, each unavailable to preceding dyads on the line, there is now a cluster of two dyads, each

---

*The arrows, numerals, words, etc. used here and elsewhere in this chapter are *not* part of the formalism, but serve only as aids to exposition.

still available for interaction with the other, as well as to occurrent dyads. This corresponds to the *second dimension*, that
of the plane, in which there are lines of interaction side by
side (as shown above) and in which there are four possible
immediately succeeding interactions, four degrees of free-
dom: (1) reciprocal interaction, as

/0/00//000/0000//   /00/0//0000/000//
/000/0000//0/00//   /0000/000//00/0//

(2) reciprocal exchange, as

/0/00//0000/000//   /0000/000//0/00//
/00/0//000/0000//   /000/0000//00/0//

in which each element's second interaction is with the reciprocal of the first, (3) switched occurrence, and (4) paired
occurrence.

4. Paired recurrence, in which the elements of a reference dyad interact with the elements of a dyad that had previously entered the history of the reference dyad, other
than that of a reciprocal exchange, that is, after an interval
of reciprocal or occurrent interactions. This corresponds to
the *third dimension*, that of volume, in which there are interactions looping outside the plane and reconnecting with it
from the outside. Here the elements have six possible kinds
of immediately succeeding interactions, six degrees of free-
dom: reciprocal interaction, reciprocal exchange, switched
occurrence, paired occurrence, switched recurrence, and
paired recurrence.

5. Switched recurrence, in which each of the two elements of the reference dyad interacts with an element of a
dyad that had previously entered its history, but in which
the two recurrent dyads are unavailable for interaction with
each other because of the intervention of the reference
dyad. This, as with switched occurrences, creates two lines
radiating out in opposite directions from the reference
dyad, but *kinked* lines, in which elements loop back to elements that had previously entered their histories, instead of
the straight lines of pure switched occurrences of the first

dimension. A switched recurrence corresponds to the *fourth dimension* — not that of time (to which I shall soon turn) but a fourth dimension of space.*

The fourth (and other extra) dimensions are not evident to sense because switched recurrences annihilate space *without re-creating it*. While, after each sequence of switched occurrences, a new line of reciprocating dyads is created (each dyad related to the others through their shared history of disturbance), a switched recurrence keeps recurring, annihilating space recurrently. Since each loop disturbance is instantaneous (as we shall show in our electromagnetic radiation model below) and since disturbances of a loop follow each other without intervals, there is no time, hence no chance, for space to reform.

Since space is composed of reciprocating dyads related to each other in their histories, the ratio of reciprocating to non-reciprocating interactions in a group, cluster, or field of elements determines the space occupied by the group, cluster, or field. The greater the proportion of reciprocating interactions, the larger the space occupied by the group, cluster, or field; the less the proportion, the smaller the space.

The interactions of a moment produce only unrelated points (if the interactions are reciprocal) or cause the annihilation of points (should the interactions be non-reciprocal). For the creation of space, in all its dimensions, sets of interactions are needed, different and changing in dimensionality, thus giving substance to space as a plenum of all the possible kinds of interactions, and the spatial dimensions to which they correspond. Time is accordingly neces-

---

*My fourth dimension may well be a manifold of extra dimensions, perhaps the six or seven posited by certain theorists, perhaps even more, all deriving from the various kinds of symmetries and a-symmetries, periodicities and a-periodicities, and the loops upon loops, of switched recurrent interactions. But I have not worked out such possible developments of dimension theory.

sary to the realization of space. Without time, there is only
the creation and annihilation of unrelated points and the
instantaneous propagation of unrelated lines, and not space
as a plenum. Time, the *nth dimension*, is the precondition for
space. There can be no space without time; accordingly
space–time.

There are basically two kinds of time: *intensive time*, the
time determined by the sequence of interactions of particular elements, as signified by strokes in element names. As
demonstrated in Chapter 3, intensive time passes at the
same rate for all elements. *Extensive time* is the time determined by the succession of interactions in a straight or
kinked line arising from switched occurrences or recurrences, from the relations accordingly of before and after
— each occurrence or recurrence in a line happening before the next one and after those that preceded it. As we
shall see in the electromagnetic radiation model, the two
kinds of time do not depend on each other. A process that
is intensively instantaneous can take an enormous time
extensively.

There are different levels of intensive time, dependent
on the level of development of the elements of group, cluster, or field for which time passes. (Level of development,
as defined in the preceding chapter, depends on the length
of a sequence of occurrences between recurrences.) The
level of development of an element at the moment of a recurrent interaction depends on the number of occurrent
reactions between the last recurrence and the present one.
It follows that the larger the number of elements mutually
interacting in a group, cluster, or field, the higher will the
average level of development in that group, cluster, or field
tend to be. For the greater the number of elements available
to each other, the wider the range of possible occurrences
and the greater their frequency.

Development presupposes recurrence. A straight-line
string of switched-pair occurrences is at the null level of development. Nor does a continuously reciprocating dyad de-

velop — for the opposite reason; no occurrences intervene between recurrences; there is continuous recurrence. There is accordingly a hierarchy of intensive times, depending on the average level of development of group, cluster, or field: from that of the undisturbed point (the continuously reciprocating dyad), for which there are no occurrences so long as it remains undisturbed, and that of a line of switched-pair occurrences, for which there are no recurrences (which, I shall posit, corresponds to an electromagnet disturbance) — in both cases, the null level of development — through neutrinos, which have the briefest sequence of occurrences between recurrences, then other leptons, then hadrons, atoms, molecules, macromolecules, the progressively higher levels of biological organization, to humankind and its social organizations, in which the ratio of occurrences to recurrences are at their highest and the average level of development correspondingly high.*

*Real time, whether intensive or extensive, would be measured by counting the number of interactions between reference events: extensive time, by the number of elements successively involved in a disturbance; intensive time, by the number of interactions involving a particular element; and the developmental level of intensive time by precise determination of the ratios of occurrent and recurrent interactions for particular sets of elements in a given duration. These procedures are, of course, impracticable. There must accordingly be alternative methods of measuring time, each with its advantages and disadvantages. Among the most important:

1. Clock measures, based on regular and dependable periodicities of reference events, e.g., the intervals between appearances of the sun at the zenith, or those between oscillations of a pendulum, or those between an atom's emissions of photons of specific frequency. Limitations of clock measures arise from departures from perfect periodicities, even in atomic clocks, in which periodicities are most regular and dependable; and the inability to indicate the quality of time, that is, its level of development. Clock time passes at the same rate for a worm and for a genius, for a stagnant society and a dynamic one.

2. Entropic measures, reckoned by the degree of randomness and homogeneity among reciprocal, occurrent, and recurrent interactions in a

Explanations

The models just presented enable us, I submit, to explain the Uncertainty Principle of Quantum Theory and the time dilation of Relativity Theory and to resolve the paradoxes related to them.

As to the Uncertainty Principle, the impossibility of precisely determining in one act of measurement such conjugate parameters as the position and momentum of a parti-

---

cluster or a field. The main limitation of this measure is — as the work especially of I. Prigogine has shown — that there are fluctuations in the process of increasing entropy and that in nonequilibrium processes, which dominate many kinds of physical and chemical systems, and are particularly characteristic of life, the tendency toward increasing entropy is reversed and, instead of increasing randomness and homogeneity, there is increasing heterogeneity and order.

3. Cosmological measures: The imputed age of the universe — 15–20 billion years are the figures most frequently cited currently — is inferred from evidence of an expanding universe — chiefly the correlation of the degree of red-shift of a galaxy's spectral lines with the dimness of its image (from which its distance is derived), from Grand Unified Theories and from much else in contemporary physics. But such inferences are questionable. For one thing, powerful assumptions are required to justify the backward extrapolation of the apparent recent and presumably current rate of expansion to the presumed moment of creation. For another, current theories assume a space–time, its dimensions and other properties, its geometry, without deriving them (as I attempt to do) from more fundamental concepts of natural order. But J. A. Wheeler, among others, has expressed dissatisfaction with assuming what has to be grounded and explained.

4. Paleontological measures inferred from the positions of fossils in rock strata, carbon dating, the counting and measuring of tree rings, and other "facts" of observation. However, such facts, and the inferences drawn from them, can be and have been sharply challenged as to their accuracy by main-stream scientists and scholars, and categorically rejected by Creationists, who also, of course, reject scientific measures of cosmological time.

5. Archaeological and historical measures derive from the study of works of art and craft, documents and other artifacts. Such inferences are continually being questioned.

6. Psychological measures depending on a person's perception of

cle: An approach to making this principle intelligible was provided by Zeno. He pointed out that a body cannot be said to have velocity, therefore momentum, when its position is precisely fixed, for at that moment it is not moving. But how, he asked, can moments of rest add up to a movement? And how could an object move a finite distance if there is an infinity of points to be traversed between any given point on a path and any other, however close? To the second question, my answer is simply the Postulate of Elementality, which denies the infinite divisibility whether of space or time, or matter or energy. As to the first question, disturbances, movements, and other spatial processes occur through the annihilation of points and the creation of new points *after* the interactions involved in disturbance, movement, etc. A particle or other body doesn't move *through* a space; it creates its space as it moves. Before the interaction resulting from a disturbance or a movement, a reciprocating dyad, a space–time point, exists in isolation, unrelated to other points, therefore not part of a space. An undisturbed point is not spatial. It becomes spatial, through relationship to other points, only *after* an interaction that transforms it to another point.

As regards the Uncertainty Principle, a particle at a point is not moving; therefore it has no momentum. When it does move, it annihilates space, creating, as a result of its movement, a new space. A particle is not at any particular point when it is in process among points. Indeed, it is not in space when moving. There is space before and space af-

---

time's passage and on recollections of past times. Such dependence is notoriously defective. It is well known that an uneventful period will "drag" during its duration, yet be remembered as inordinately brief; while an eventful period will pass quickly but be recollected as having lasted a rather long time. An "absent-minded" person, and, more extremely, one suffering from *petit mal*, will take up what he or she was doing before the onset of his inattention or spell, with no awareness of the time lapse, indeed often vehemently denying that any occurred. Also, drugs variously affect time perception.

ter the interactions — non-reciprocal, therefore not spatial — but not at the moments of interactions. Where space
is, momentum is not. When momentum is, space is not,
since movement and other disturbances annihilate space.*
In attempting to determine a particle's position, electromagnetic energy must be employed, thereby energizing,
disturbing the space enclosed in and enveloping the particle, annihilating that space and thus rendering spatial position indeterminate. So, too, in attempting to determine a
particle's momentum at a certain position, the energy employed in the determination raises the particle's energy
level, therefore its momentum, making the momentum at
the moment of measurement indeterminate. The combined
indeterminism of position and momentum is proportional
to the frequency and intensity of the radiation used in the
attempted determination, with ħ the presumed unit of disturbance, as the lower limit. *Neither* the position *nor* the momentum of a particle can be known precisely, at the moment of attempted determination, for the determination
destroys the position and alters the momentum. They *can*
be known precisely, but only *after* the attempted determination, and then only inferentially — by measuring the particle tracks photographed in a bubble chamber, by counting
clicks in a counter, or in other appropriate ways.†

As to relativistic time dilation, that is of two kinds, both
explained in different ways, only the second of which per-
tains to the conceptual framework here presented.

First, apparent time dilation: According to Special Rel-

---

*Incidentally, Einstein, though uncomfortable with indeterminacy,
showed that the more sharply a system's time is determined, the less
sharply is the system's energy at that time determinable. For, at an instant
of time, there is no energy, which entails a frequency, a succession of
disturbances, hence a duration.

†It should be noted that the Uncertainty Principle, and the indeterminacy that is its consequence, restricts what can be *measured* of a particular
particle state at a given moment, not the way the state evolves. That remains deterministic, though the determinism is probabilistic.

ativity Theory, as a spaceship recedes in uniform motion from an earthbound observer, light or, more precisely, the information carried by light signals about clock readings, takes longer to reach the observer, because the path traversed by the signals continuously lengthens. Therefore to the earthbound observer, the spaceship's clock appears to be running slow. To the spaceship navigator, the clock of the observer on earth also appears to be slowing down, for the same reason. Apparent time dilation is symmetrical.

As to real-time dilation: According to General Relativity Theory, there is real-time dilation with accelerated motion (change in velocity or direction or both). This has been verified experimentally. Thus, the half-life of a mu-meson will increase appreciably as it accelerates toward the speed of light. An astronaut rushing about in space (changing speed and direction) will age less rapidly than his twin on earth. The reason: as a body (whether a sub-atomic particle or a huge spaceship with human occupants) accelerates, the proportion of spatial occurrences involving the body's constituent elements increases, while the proportion of mutual interactions among the body's elements correspondingly decreases. Differently put, there is a larger proportion of interactions at the spatial level of development, a smaller proportion at the physical and biological ones. But aging is a function of interactions at the physical and biological levels. Such slowing down of time, and particularly of the aging process, occurs in common experience. A hibernating or cold-blooded animal will experience a slowing of its biological processes as the temperatures to which it is exposed drops. So, too, for a refrigerated person, biological time would pass more slowly, and so too would the aging process — at the extreme cold of interstellar space, hardly at all. (Hence the plausibility of extending a person's lifespan indefinitely by hurling him into space, to be returned to earth at any time that earthlings may decide, perhaps centuries or millenia later.) For a hyperactive person rushing about all the time, taking little of his time to integrate his experiences in feeling, reflection, and thought (which depend on a high proportion of excitations among the inter-neurons, those in the the brain and especially the cerebral cortex) emotional and intellectual maturation would tend to proceed more slowly than in a more normal person.

The Electromagnetic Radiation Model

Let us suppose a linear arrangement of three space–time points (that is, reciprocating dyads). By linear is meant an arrangement in which one dyad is between the other two and available for interaction with both, while the two end dyads are unavailable for interaction with each other at the instant of their availability to the middle dyad, the one intervening between the two others. The following represents such a situation (the participating elements being metalinguistically labeled):

(1) /0/00/   (3) /000/0000/   (5) /00000/000000/
(2) /00/0/   (4) /0000/000/   (6) /000000/00000/

Consider now the consequences of the interaction of 1 and 3 and of 4 and 6. This would leave elements 2 and 5 unengaged. Since they are assumed, by the nature of the posited arrangement, to be unavailable to each other in the first instant, they could not, at that instant, interact with each other. Nor could they, by the Corollary of Dyadal Interaction, become involved as third elements in the other two interactions. Furthermore, it follows from that Corollary that neither element 2 nor element 5 could remain, even for one moment, uninvolved in some elemental interaction. Since it has been stipulated that neither of these two elements is available to the other at the reference instant, and their reciprocals of the immediately preceding instant being otherwise engaged, it follows that both of these odd elements must, at the instant of disruption of the dyads of which they had been members, interact with other elements outside the original set of six.

Let us say that element 2, i.e., /00/0/, interacts with one of the elements of the reciprocating pair 7, /0000000/00000000/, and 8, /00000000/0000000/, while odd element 5, i.e., /00000/000000/, interacts with one of the elements of the reciprocating dyad 9, /000000000/0000000000/, and 10, /0000000000/000000000/. This would leave the other members of the two occurrent pairs, either 7 or 8 and either 9 or 10, unaccounted for. *These* odd elements must also, without a moment's delay (since no instant can pass without an interaction) interact with members of other occurrent dyads; and so on indefinitely. Thus more and more dyads will become involved in a series of *split-pair occurrences,* constituting *a switched-pairs disturbance.*

Thus, the splitting of a dyad, a space–time point, causes a disturbance radiating out in one dimension but in opposite directions from the split point. The disturbance is instantaneous at each point and intensively instantaneous also all along the line; that is, the disturbance at each point is, in the intensive dimension of time, neither earlier nor later than that at any other point — since at the same instant that a dyad is split the odd elements must engage elements of available dyads, causing them to split. There is accordingly no lapse of intensive time between the splitting of dyads along a line, no matter how many dyads may be involved, however long the line. However, at the moment of disturbance, there is *no* line nor indeed any points that might constitute a line; the disturbance annihilates the points involved. The distance between any points on the line is accordingly zero. A line is created as the *result* of the disturbance, because the new points created by the reciprocal interaction of the newly constituted dyads are now related to each other, having entered each other's history, and through such relation come to constitute a line, a set of connected points.

But while there is no lapse of intensive time during a disturbance, there is lapse of *extensive* time, dependent on the succession of interactions between the initial splitting of a dyad and all subsequent splits: the first splitting of a dyad,

then a second, then a third, etc. Thus, while the whole disturbance is *instantaneous*, the times along the line of disturbance are all different — the times are *not* simultaneous.

Why, it may be asked, if there is a before and an after
along the line of disturbance, should not successive elements along the line have added strokes in their name, the
number of strokes in each name dependent on relative position in the succession of splits? The answer is that the elements up the line (nearer the point of initial disturbance)
do not enter the history of those further down. Only the
immediately preceding odd element, the one causing the
splitting of a dyad, enters the history of one of the elements
of that dyad. Consider the disturbance

$$(1) \quad /0/00/ \longleftrightarrow (2) \quad /000/0000/$$
$$(4) \quad /00000/000000/ \longleftrightarrow (6) \quad /0000000/00000000/$$
$$(3) \quad /0000/000/ \longleftrightarrow (5) \quad /000000/00000/$$

Because of the disturbance, in the *next* instant we have the
line

$$/0/00//000/0000// \qquad /0000/000//000000/00000//$$
$$/00000/000000//0000000/00000000//$$

together with their reciprocals /000/0000//0/00//, etc.

We can see that the history of 1 has entered into that of
2, and the history of 3 into that of 5, etc.; but the history of
2 has *not* entered into that of 5, nor the history of 3 into
that of 6. Thus the line created is connected only point-to-
point, the one immediately preceding to the one immediately following. Points more remote are not connected —
except by inference, which intellectually synthesizes a
succession of connected points into a connected line. The
extensive time of a disturbance, determined — though, of
course, quite beyond possibility of empirical determination
— by the number of points disturbed between two reference points, can be, and has with increasing precision been,
empirically measured. From that, the velocity of light, or
other electromagnetic radiation, is derived.

I submit that the split-dyad, extending to a switched-pair disturbance, corresponds to, essentially models, an electromagnetic disturbance, each such disturbance presumably having the value ℏ, the quantum of electromagnetic action — energy multiplied by time. This unit of action is a constant, with the same magnitude for all observers, whatever the differences in time *or* energy considered separately.

A *photon* would then be a succession of disturbances in one space dimension, having a certain characteristic frequency of disturbance, which determines its energy. Each of the lines created by disturbance is part of a manifold composed of all the disturbed lines between the source of the disturbance and the receiver. The lines are not the same, nor can they be said to be parallel. Rather are they, for each disturbance, reconstituted lines, reconstituted *after* each disturbance, since (as already indicated) a disturbance annihilates space, occupies no space and therefore has no linear (or other) dimension during the instant of disturbance. Since the source of disturbance (an atom or sub-atomic particle) is extended in space, the lines formed by a set of disturbances from the source will not coincide. A photon is accordingly a bundle of disturbances with the same (extended) source and extended receiver (also an atom or sub-atomic particle). The bundle may have either a random distribution of disturbances (and of the lines they form) and be therefore unpolarized, or have a particular distribution, configuration, of disturbances and be therefore polarized (in a planar, circular, elliptical, or other mode).

Because disturbances are instantaneous, the occurrence of a photon depends on the coincidence, *at both ends*, of the disturbances that constitute the photon. The receiver "causes" the disturbance, no less so than the source. Source and receiver must resonate, in frequency, phase, polarity, for a photon to "travel" from one to the other. A photon exists only if there is not only an emitter but also a receiver ready to absorb it, thus raising its energy level. (There is only emission and absorption. Reflection and refraction are the consequence of concatenated sequences of absorptions,

new emissions, new absorptions, etc., among the clusters of
atoms and sub-atomic particles involved.) It is only in such
optical phenomena that photons manifest themselves, indeed exist at all. (For a source disturbing reciprocating
dyads in all directions, the incidence of resonances among
particular receivers diminishes with the square of the distance from the source.)*

This is the reason for the wave–particle dualism of a
photon's nature, and of that of all sub-atomic particles. A
photon or material particle becomes manifest only in its effect on other particles. Otherwise there are only potentiali-
ties, known only inferentially, for creating such effects. But
for the effect to occur, the receiving particle must be recep-
tive; it must resonate, in its periods, charge, spin, polarity,
and other properties, with the impacting particle. But there
is no certainty, only a certain probability, that this will be the
case. Thus, do probabilities, as represented by wave functions,† propagate through space (in the direction of the particle's movement). The intensity of the field, the height of
the wave, corresponds to the degree of probability that resonance will occur. When it does occur, the wave function
"collapses" to a definite event, the absorption of a photon,
the impact of a material particle. In quantum mechanics, a
particle is represented by a wave packet, which is the
consequence of constructive and destructive interference
among wave trains extending, theoretically, all through
space. What "travels" are the nodes of the waves in the

---

*Compare a nervous excitation: A neuron "fires" only if conditions all
along its axon, cell body, dendrites, and the synapses at the ends of axon
and dendrites are in resonance. Only if all conditions — especially those
at the synapses — are right will there be an excitation. If one of them is
wrong, there will be none. Nervous excitations are accordingly subject to
an "all-or-nothing" law, as electromagnetic disturbances are.

† See E. P. Wigner, in *Symmetries and Reflections* (Bloomington, Indiana
University Press, 1967): "Given any object, all the possible knowledge
concerning the object can be given as its wave function. . . . More precisely, the wave function permits us to foretell with what probabilities the

packet. With observation or other registration (that is, absorption or impact), the wave packet collapses.

A photon's or a material particle's wave aspect corresponds to the probability that conditions at both emitting and receiving ends resonate. The particle aspect corresponds to the collapse of the wave when conditions do actually resonate and propagation actually occurs, however improbable the occurrence may previously have been calculated to be.

For two photons propagating between a particular emitter and a particular receiver, the probabilities of propagation can either constructively or destructively interfere with each other, that is, tend either to re-enforce or to cancel each other out. So much accords with classical optics. Where quantum theory differs is in the demonstration (as in Young's famous two-slit experiment) that a *single* photon can interfere with itself. The reason is that a photon is not a single disturbance but a sequence of disturbances along (rather creating) different lines between source and receiver. There are no certainties, only probabilities, that source and receiver will resonate along particular lines of disturbance. The probabilities will differ for different lines. Since each receiver — always a sub-atomic particle or an atom composed of such particles — extends in space, the probabilities will vary along a wave front parallel to the receiving body. At some points along this front, the probabilities associated with particular lines of disturbance will re-enforce those associated with other lines, and a photon, a sequence of disturbances having a certain frequency, will

---

object will make one or another impression on us if we let it interact with us either directly or indirectly. The object may be a radiation field, and its wave function will tell us with what probability we shall see a flash if we put our eyes to certain points, with what probability it will leave a spot on a photographic plate if this is placed at certain positions. In many cases, the probability for one definite sensation will be so high that it amounts to a certainty," which is the case, as Wigner goes on to say, with classical mechanics but not in quantum mechanics.

manifest itself. At other points, the probabilities will cancel each other out. There will still be disturbances at these points, but no resonance in frequency, polarity, etc. that characterize a photon and that are preconditions for its existence.

Because light (and other electromagnetic radiation) doesn't traverse space but rather annihilates it, it has the same "velocity" for all observers, whatever their speed and direction of motion, or their acceleration. There is no space, *at the moment of propagation*, between the source of the radiation and the observer, whether near or far from the source, whether speeding toward it or away from it or across the line of propagation. (As explained below, electromagnetic radiation actually has different "velocities" in different media. However, in the same medium, whatever it may be, it is the same for all observers.) A signal will take less *extensive* time to reach an observer as he moves toward the source, because fewer points are disturbed as the signal is propagated — there are fewer different times — the shorter the path of the propagation; but the propagation is everywhere, including the observer's eye or instrument, instantaneous.

Because a split-pair disturbance is a string of occurrences, and only of occurrences, it has the greatest possible "velocity." That is, its extensive time is the minimum possible because it is all occurrence, no recurrence, no "lingering," as it were, along its path. Because recurrence slows down a disturbance, the "velocity" of electromagnetic radiation is greatest in a pure (that is, particle-free) vacuum. In other media, which contain particles, therefore recurrences, odd-element disturbances may get trapped, either to be absorbed, thus raising the energy level of the receiving particle, or to cause new disturbances in processes of reflection and refraction. The greater the proportion of such recurrences, the lower the "velocity." Because all media, even the densest, are preponderantly empty space, the slow-down is slight.

Because, as I shall soon attempt to show, matter tends to annihilate space in its vicinity, its field, the "speed" of light also tends to increase in a gravitational field, the increase being greater, the stronger the field. A photon's frequency also increases (a blue shift), as the field's intensity rises, there being less space to disturb in successive odd-element disturbances.

Thus the speed of light is a constant only in a particle-free and gravitation-free vacuum, an unattainable ideal. To the extent of departures from this ideal — in actuality, slight, except in the vicinity of extremely massive stars, and, at the extreme, black holes — there is departure from the constant.

Movement through space means occurrences with the points that constitute a given space, converting reciprocating dyads into non-reciprocating pairs, either split or paired, therefore either into energy or matter (bound energy, as we shall see). Because of this conversion, the mass of a moving body, the amount of energy–matter it contains, increases. In Relativity Theory, as a moving body approaches the speed of light, its mass would tend to infinity. However, matter, unlike energy, requires recurrent interactions, however remote from each other they may be. But light is all occurrence, no recurrence, therefore no matter. It follows that, at the limit of the speed of light, a moving body would be totally converted to unbound energy.

That odd-elements cannot interact with each other, and, by forming a dyad, end a disturbance, follows from the Corollary of Dyadal Interaction. Elements exist only in pairs, dyads. An odd element, crossing the line of disturbance, other than that of which it is a link, could only encounter dyads, which it would split, thus continuing its own disturbance. Disturbances can, and do, cross each other; but they do not, cannot, negate each other; because all along the lines of the disturbances, there are only dyads — switched pairs, to be sure, each element in the line having new partners, but still dyads, no odd elements. Therefore, a disturbance is forever, in extensive time. It may remain linear, extending back therefore to the Big Bang and forever to the Big Crunch or however the universe ends (if indeed it does ever end) or it may become trapped in a multi-dimensional particle, bound electromagnetic energy; but it cannot be neutralized or otherwise negated.

Nor can new disturbances occur. For the sake of exposition, the disturbance model given above was represented as the splitting of a dyad. But this is misleading. There is *no* splitting of dyads, only the switching of elements among dyads, each switch in a line of disturbance being one of an unbroken series of switches. Consider again the initial model (metalinguistically abridged), in which 5 and 6 constitute the reference dyad:

$$1 \longleftrightarrow 3 \qquad 5 \longleftrightarrow 7 \qquad 9 \longleftrightarrow$$
$$\longleftrightarrow 2 \qquad 4 \longleftrightarrow 6 \qquad 8 \longleftrightarrow 10$$

As is evident, the dyad 5 and 6 differs in no way from that of the other dyads in the line of disturbance. Each switch "causes" the one next in line in both directions. The situation is like that of two rods lying parallel to each other. The shift of one unit of length in one of the rods relative to the other (while remaining parallel to it) is *not* caused by any point in the displaced rod. The cause of the switch in the relationship of points in the two rods — if cause there be — must lie outside the rods.

However, by the Constraint of Sufficient Reason, we cannot postulate an outside cause for the switching of elements in a line of dyads. For, by the Postulate of Elementality, there is nothing except elements; and, by the Corollary of Dyadal Interaction, all elements always occur in pairs. No action by a dyad upon a line of dyads can cause a disturbance of the line. The "intruding" dyad simply becomes one of the switched dyads of the line of disturbance. One may of course postulate a divine intervention— *fiat lux* — as the outside cause of a disturbance, but not within the conceptual framework here adumbrated. Nor, by the Constraint of Sufficient Reason, may we posit a spontaneous splitting of a dyad. For there is nothing we know about a dyad and its constituent elements, or about the elements with which they may interact, that would permit us to say that a split may spontaneously occur.

The conclusion is, I submit, inescapable that electromagnetic energy can neither be destroyed or created; it is accordingly conserved. From the conservation of electromagnetic energy, the conservation of matter–energy (disturbances whether linear or bound) follows, and so probably also the other conservation laws. But these derivations will not be attempted here.

Explanations

This section is perhaps the most critical in this work. Upon its effectiveness in meeting the challenge of intelligibility presented in chapter 1 and here, somewhat more sharply, reiterated, the work's value largely depends. Einstein tried to meet that challenge — most notably in his 1935 paper written with B. Podolsky and W. Rosen, "Can quantum mechanical description of physical reality be considered complete" (*Physical Review* **47**, 1935). Though he kept trying until his death to overturn the refutation of his argument by Bohr and others, the almost universally accepted verdict is that he failed. So, too (in the current opinion of the great majority of theoretical physicists and philosophers of science), did the efforts of de Broglie, Vigier, Bohm, and other "hidden-variable" theorists fail. Now the overwhelming majority of practicing physicists have stopped trying to meet the challenge. They admit, in effect — some fiercely and dogmatically assert — that physics, and the reality it seeks to describe, is fundamentally unintelligible; and they then confine themselves to observation, measurement, and the calculation and prediction of future observations and measurements. The most imaginative and daring ones seek to unify their measurements and calculations in ever more comprehensive sets of equations, Grand Unifying Theories (GUTs) and Theories of Everything (TOEs) in which all of reality is rendered basically *in*comprehensible on any terms acceptable to common sense.

What has led physicists to abandon the quest for intelligibility, and what has accordingly still to be explained, is the mounting evidence for what has been termed "nonlocality" or "nonseparability," which, in essence, means that two photons — or other sub-atomic particles, though the extension of the concept to material particles is more questionable than its application to photons — that have a common source but that propagate in different directions, and that have not interacted since emission with any other system, must be considered to constitute a "nonseparable" system, so that what happens to one photon will affect the state of the other, no matter what the distance between the locations at which the two photons are absorbed and thus detected, even though an influence of one absorbed photon upon the state of the other would have to propagate at velocities exceeding that of light, which physicists universally regard as impossible. In terms of quantum theory, this corresponds to the assertion that it is not permissible to associate with either of the photons, as characterized above, its own wave function. A wave function can be attributed only to the composite system, though its two constituents, at the times of absorption, be billions of light years apart. This means that, under the conditions described, what happens to one photon will determine what happens to the other, though no possible causative influence is known, or can be conjectured without violating the basic exclusion principle of Relativity Theory: that no causative influence can travel faster than the speed of light.

The most compelling evidence for this thesis is that derived from two sets of observations, made or refined within the last few years:

1. Delayed choice: This can most conveniently be described by reproducing an illustration from J. A. Wheeler's paper read at a joint meeting of the American Philosophical Society and the Royal Society, on June 5th 1980:

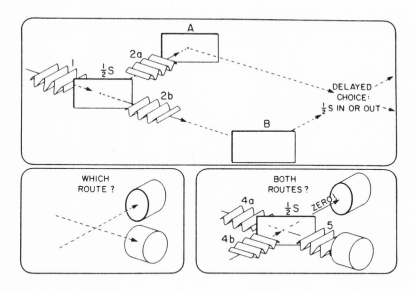

Fig. 1. Beam splitter (above) and its use in a delayed-choice experiment (below). An electromagnetic wave comes in at 1 and encounters the half-silvered mirror marked "½S" which splits it into two beams, 2a and 2b, of equal intensity which are reflected by mirrors A and B to a crossing point at the right. Counters (lower left) located past the point of crossing tell by which route an arriving photon has come. In the alternative arrangement at the lower right, a half-silvered mirror is inserted at the point of crossing. On one side it brings beams 4a and 4b into destructive interference, so that the counter located on that side never registers anything. On the other side the beams are brought into constructive interference to reconstitute a beam, 5, of the original strength, 1. Every photon that enters at 1 is registered in that second counter in the idealized case of perfect mirrors and 100 per cent photodetector efficiency. In the one arrangement (lower left) one finds out by *which* route the photon came. In the other arrangement (lower right) one has evidence that the arriving photon came by both routes. In the new "delayed-choice" version of the experiment one decides whether to put in the half-silvered mirror or take it out at the very last minute. Thus one decides whether the photon "shall have come by one route, or by both routes" after it has "*already done*" its travel."

Wheeler sharpens his point on the following page: "There we make the decision whether to put the final half-silvered mirror in place or to take it out in the very last picosecond, after the photon has already accomplished its travel. In this sense, we have a strange inversion of the normal order of time. We, now, by moving the mirror in or out have an unavoidable effect about what we have a right to say about the *already* past history of that photon." (his emphasis).

In the same paper, Wheeler describes such observations on a "cosmological scale":

Two astronomical objects, known as 0957 + 561 A, B, once considered to be two distinct quasi-stellar objects or "quasars" because they are separated by six seconds of arc, are considered now by many observers to be two distinct images of one quasar. Evidence has been found for an intervening galaxy, roughly one quarter of the way between us and the quasar. Calculations indicate that a normal galaxy at such a distance has the power to take two light rays, spread apart by fifty thousand light years on their way out from the quasar, and bring them back together at the Earth. This circumstance, and evidence for a new case of gravitational lensing, make it reasonable to promote the split-beam experiment in the delayed-choice version from the laboratory level to the cosmological scale. . . . We get up in the morning and spend the day in meditation whether to observe by "which route" or to observe interference by "both routes." When night comes and the telescope is at last usable we leave the half-silvered mirror out or put it in, according to our choice. The monochromatizing filter placed over the telescope makes the counting rate low. We may have to wait an hour for the first photon. When it triggers a counter, we discover "by which route" it came with the one arrangement; or, by the other, what the relative phase is of the waves associated with the passage of the photon from source to receptor "by both routes" — perhaps 50,000 light years apart as they pass the lensing galaxy G-1. But the photon has already *passed* that galaxy billions of years before we made our decision.

This is the sense in which, in a loose way of speaking, we decide what the photon *shall have done* after it has *already* done it" (Wheeler's emphasis).

2. Violation of Bell's Inequality: Even more striking are the experiments made by the French experimental physicist A. Aspect and his associates,* and confirmed by other experimenters. So astonishing are the apparent implications of these experiments that even physicists who eschewed the quest for intelligibility have had to pay attention.

In essence, the experiments were designed to determine the effect on the polarization of two photons having a common source but propagating in opposite direction by altering the angle of a calcite polarization detector at the end of one's photon's path *relative to* the angle of the polarization detector at the end of the twin photon's path, when the two ends can be arbitrarily far apart. The experimental results seem conclusively to show that the "effect" on the photon's polarization at detector A correlates *more strongly* with the polarization at detector B when the relative angle at B is changed than simple and indisputable correlation analysis allows. This implies that the states of the two separated photons are more tightly interdependent than would be possible were two photons truly separated. But that directly contradicts Einstein's affirmation (in "Autobiographical Notes," Schilpp, *op. cit.*): "On one supposition we should, in my opinion, absolutely hold fast: the real factual situation of the system (G) is independent of what is done with the system (B) which is spatially separated from the former."

The explanation I offer follows directly and simply from the distinction between intrinsic and extrinsic time in

*Reported by him and by his co-authors P. Grangier and G. Roger in a 1981 paper, "Experimental tests of realistic local theories via Bell's theorem" (*Physical Review Letters* **47**), which I have seen, and by later papers, which I have not.

my electro-magnetic model. Electromagnetic disturbances are intensively instantaneous, no matter how long the line of disturbance. There is *no* time lapse. What happens, what is done, at either end of a line of disturbance (in the delayed choice, the violation of Bell's inequality, or related experiments) *instantaneously* affects what happens at the source of the disturbance and at the other end of the line. The cause of what happens spreads out instantaneously over source and both (indeed all) lines of disturbance.

But while the disturbance is instantaneous, the instants at the different points in the line, are different; they differ in the order of succession. The disturbances at different points are *not* simultaneous. The dyadal switches close to the source of a disturbance occur *before* those more distant from the source; those more distant occur *after* those closer. There is accordingly lapse of *extensive* time. "Seen" from outside the line, there is no before and after, but only an instantaneous disturbance. But, of course, a disturbance cannot be *seen* outside the line of disturbance. That is why there is no possibility of communication at superluminal velocities (contrary to parapsychologists and certain of the "hidden variable" theorists). Cause, as determined by electromagnetic disturbances, is instantaneous; but communication cannot be faster than light. The first is a function of intensive time, the second of extensive.*

Thus the act of observation *does* cause the event observed, since the event and the observation constitute the

---

*Whether or not the violation of Bell's Inequality, and accordingly instantaneous causation, applies to material particles, as well as to photons, is an open question. It has been argued that it does, particularly to the spin of particles once connected. There is (to my knowledge) no clear experimental demonstration of this effect. If it does exist, the reason can only be (in the context of this work) that spin must involve linear disturbances, intensively instantaneous, though not necessarily with the frequencies, polarities, and other attributes of electromagnetic radiation. But my models of material particles are not yet complete; and therefore I cannot now show how such influences might propagate.

same physical phenomenon, instantaneous for each of the successive instants of the phenomenon. A photon is transmitted, it occurs, when the observer's state, or that of his registering apparatus, resonates with that of the photon's source. What the observer does, how he determines his state, or that of his apparatus, determines which of the lines of disturbance between source and receiver fulfills its potential of becoming a photon, with its characteristic frequency, polarity, etc.

The extrinsic identity of an element, and therefore of a subatomic particle (which is hypothesized as composed of elements) indeed the character of any physical object, depends on the history of its internal and external (field) interactions, including therefore the history of observations, measurements, and other manipulations made upon it. Each such event affects its history, therefore its extrinsic identity, its character. While all physical objects are susceptible to external disturbance, including those caused by acts of observation, measurement and other manipulation, the objects of micro-physics, the sub-atomic particles, are especially susceptible, because the photons that are indispensable to observation are close in magnitude to the objects observed. An act of observation — always a two-way act, going from observer to what is observed as well as from the object observed to the observer, and doing so instantaneously — becomes a significant part of the sub-atomic particle's history, therefore its extrinsic identity. The act of observation *doesn't* create the objective reality, but by his act of observation, the observer helps determine it. What he particularly determines are the aspects of reality most closely related to the propagation of photons, therefore a particle's position, which can only be determined by photons propagating between particle and observer; and momentum, which is altered by such propagation, each of which raises the particle's energy level. Those aspects of a particle's character, its rest-mass, its charge, its spin, etc., that are not affected by the incidence of electromagnetic radiation between itself and an observer are, for that very reason, not affected by the act of observation. A macroscopic body in which internal interactions overwhelmingly predominate over those involving propagation of photons between itself and observers is little, if at all, affected by acts of observation involving photons (unless the radiation is of such intensity as to raise the body's energy level to a damaging or destructive level).

Gravitational Models

I propose that matter consists of captured odd-element disturbances, bound energy. While an electromagnetic disturbance is instantaneous, because purely occurrent, and linear, bound disturbances are recurrent — the lines cross, therefore intensive time elapses — and in four or more dimensions. There are occurrences, split and paired, and recurrences, split and paired. Each recurrence, each crossing of a line (to form the loops, kinks, twists, knots, etc., that the different species of particles exhibit) is later than the one before. The succession of crossings determines the duration, which, for stable particles (see below) may approach eternity.

Non-reciprocal interactions are non-spatial — they negate space. The ratio of reciprocal to non-reciprocal interactions in a particle or in a body (an aggregation of spatially distributed particles) determines the space occupied, or rather contained, in the particle or body. The greater the proportion of non-reciprocal to reciprocal interactions, the more dense, compact, the matter.

Particles and bodies interact also with their fields, and such interactions, being non-reciprocal, "devour" the space in the field. The annihilation of spatial fields by interaction with matter is the reason for the mutual approximation ("attraction") of particles and bodies, as a function of their respective masses (that is, of the number of elements in an aggregation of elements developed to the matter level) and of the distance of the bodies to each other (that is, the number of reciprocating dyads, space points, between particles or bodies). The greater the number of matter elements, the more numerous their interactions with space elements, and therefore the greater the incidence of space annihilations and the stronger the gravitational field. The smaller the distance between bodies, the fewer intervening space elements will there be, therefore the greater the proportion of non-reciprocating interactions in the field separating the bodies, and therefore the stronger is the gravitational attraction. As matter comes more and more to preponderate over space, there is proportionately an increasing rate of space development, therefore acceleration of the movement of the bodies toward each other.

Because the closer that space elements are to a body, the greater the likelihood of their non-reciprocal interaction with that body's elements and therefore the negation of their spatial character, space "curves" around a body (as water curves around a heating unit that vaporizes it). Or, in other terms, the space in the vicinity of a body is "warped." That is why light in the vicinity of a star or a galaxy is "bent."

A gravitational field itself has mass in proportion to the probability that the space points surrounding a body, and available for interaction with its elements, themselves develop through such interaction into matter. Matter production is greatest where the gravitational field is strongest, and therefore space curvature is greatest. The mass of a body, accordingly, never has a sharp boundary. It extends, shades into, the enveloping space in proportion to the strength of its gravitational field, and the effect of that in developing, therefore curving or warping, space.

While the strength of a body's gravitational field depends on its massiveness, the ratio of a body's internal interactions to its field interactions is the same for all bodies, whatever their differences in mass. For, the more massive a body, the greater the number of other of its elements available to each of the body's elements. So that while the more massive a body, the greater proportionately the number of its field interactions; but also, and in the same proportion, the number of internal interactions. That is why, as Galileo discovered, heavy bodies do not fall more speedily to earth than light bodies (air resistance aside). Hence, the constant of gravitation.*

On the equivalence of gravitational and inertial fields: Inertia, which is the disposition of bodies to remain at rest or in uniform motion in the same direction unless acted upon by some external force, which alters its velocity or direction or both, involves the development of the space elements in the bodies' fields, and therefore the annihilation of space in the fields, at rates characteristic of the bodies' velocity — the rate being greatest in the direction of movement. But gravitation also involves the development of space elements in the fields enveloping bodies — the rate of such space development being dependent on the masses of the gravitating bodies and the distance separating them, the rate being greatest in the direction of the gravitational "attraction." However, inertia characterizes single bodies, while gravitation involves mutual attraction among bodies. The lessening of distance means the progressive lessening of the proportion of intervening field elements to the inter-

---

*Recent experiments have cast doubt on the validity of Galileo's principle, and therefore on the constancy of the Gravitational Constant. That constant is now thought by some physicists to vary for bodies of different chemical, and perhaps also structural, composition, through the operation of a posited fifth basic force, which has been tentatively labeled "hypercharge." The evidence and arguments for the existence of this force are still, at this writing, inconclusive. Should it exist, it would reflect the differential effect on field interactions of different configurations of matter.

nal elements of the gravitating bodies, and therefore the tendency for the acceleration of motion toward each other.*

Notes for Particle Models

I propose that subatomic particles are all trapped switched-dyad disturbances and, as such, bound electromagnetic energy. Instead of the disturbances extending linearly from point of origin to point of absorption, the elements constituting a particle form a cluster (if a-periodic) or group (if periodic) closing in upon itself in fluctuant loops, kinks, twists, knots, coils, braids, and other arrays.† The crossing of lines of disturbances is the basis for intensive time, for at the nodes of crossings the histories of a disturbance's elements intersect; they enter into each other's histories to constitute an interval, a duration, the length of which depends on the number of occurrences between nodal recurrences. There was no intensive time before the closing of lines of disturbance upon themselves and then the crossing of the closed lines, to constitute matter.

The dimensions of a particle are functions of the number of reciprocating dyads enclosed within the particle's loops, knots, and other configurations, the relative frequency of paired occurrences, switched occurrences, and switched recurrences among the particle's elements. The

---

*In this model, gravitons are excluded, as is Newtonian action at a distance. Gravitational waves (still to be decisively detected) would reflect disturbances in the fields separating bodies resulting from an enormous change in mass, as when a star explodes.

†See Atkins (*op. cit.*): "Different species of particles are different species of knots in the structure of space–time. Different knots are different groupings (like everyday knots are different ravellings of string) of the binary entities coming into existence at the creation. Different particles are therefore different topological structures of space–time."

lower the proportion of reciprocating to nonreciprocating dyads, the less space does the particle enclose. At the extreme, with vanishingly few reciprocating interactions, the particle becomes pointlike.

A particle's velocity through space depends on the ratio of space occurrences, that is, interactions with reciprocating dyads, to occurrences along the lines of disturbances, recurrences at crossing nodes, and recurrent interactions among enclosed dyads. The higher the proportion of space occurrences, the greater the particle's velocity.

Particles interact when, at the nodes, there are switched pairs (field) occurrences that resonate with those of other particles, in frequency, phase, polarity, etc. There are characteristic probabilities that the disturbances characterizing particular particles will resonate. Hence the mutual attraction of oppositely charged particles and the mutual repulsion of particles bearing the same charge.

Certain configurations of closed disturbances are stable or elastic, that is, capable of restoring themselves after disturbances. They thus correspond to the stable strings and knots of topology.* The great majority of possible configurations are variously unstable, inelastic, and subject therefore to enodation, the unraveling of the knots, therefore disintegration or other transformations. These correspond to the host of unstable particles and virtual particles. It is probably the case that particles characterized by periodic recurrences at their nodes are stable, while those characterized by aperiodic recurrences are unstable.

I have been unable so far to model any of the subatomic particles — and the task would appear to be extremely formidable, even for those very much better equipped for it than I, given the number of elements and interactions even

---

*Compare also a flame or a wave in water, in which form is preserved for greater or lesser periods — at one extreme for only two moments, at the other, forever, while the substance consumed in the flame or agitated by the wave changes continually.

for an electron, let alone for such monsters as the W and Z bosons, with masses equivalent to about 10 billion electron volts.

Only with one particle, and that by far the least massive, the neutrino, have I made a tentative start toward modeling. I submit that a neutrino corresponds to a switched recurrence — a looped disturbance, therefore in more than one dimension (my proposed fourth), unlike the switched occurrence resulting in photons, which exist in only one linear dimension. Because a neutrino is only one loop or kink in space–time, it has only one node (a crossing of the line of disturbance): therefore its movement through space occupies the least possible duration (intensive time), though greater than that of a photon, which, when unbound, has no duration. Its "velocity" is accordingly only slightly less than that of a photon. It also, for the same reason, has the least mass of any particle (other than the photon). Like switched occurrences, switched recurrences are conserved. They can neither be created nor negated — though they can be unbound from larger subatomic particles, as in the conversion of a neutron to a proton, electron, and neutrino. Because they are electrically neutral and have miniscule mass, they are hardly subject to gravitational force and not at all to electromagnetic force — a disturbance is either straight or kinked, and a kink cannot capture a straight-line disturbance — they rarely interact with other particles. (They can pass through as much as a million miles of lead without interaction.) Because neutrinos exist in only one dimension (which I have called the fourth) they possess only one kind of spin.*

---

*Why the neutrino's spin should be "left-handed," when other leptons, and also photons, can have either left- or right-handed spin, I don't know. I also don't know why there should be three kinds of neutrinos, those associated with the electron, muon, and tau muon, nor just how they differ, nor how they can transmute to each other. These are only some of the questions to which, at present, I have no answers.

Hints for Cosmological Models

At the "beginning," there were only unrelated reciprocating dyads. There was nothing in their histories to indicate the existence of any other element. Each dyad was self-contained, isolated. There was no extensive time, only intensive time, proceeding separately for each dyad; no space; no energy; no matter. No universe.

Then there occurred a disturbance, a switching of pairs. (I have no idea of what caused the disturbance. The religious-minded are free to assume that it was God's *fiat lux*.) The disturbance extended through all the elements of the initial universe; in effect, it created that universe. For *one* instant, that of the instantaneous disturbance, there was no space, only electromagnetic energy. The universe was accordingly at the highest possible energy level, since it was all energy and extensive time, the succession of instantaneous disturbances through all the dyads disturbed.

Space, a relationship among reciprocating dyads, space points, came into being in the instant *after* the initial disturbance.

Then, as I conceive it, there occurred the crossing, the looping, kinking, knotting, braiding . . . of disturbance lines, the creation of matter and anti-matter, perhaps involving all the elements of the initial universe.

Then, the enoding, the unraveling of the knots — the mutual annihilation of matter and antimatter to form new linear disturbances, radiating out from the central mass, leaving a matter residue. (According to prevalent conceptions, there are between one and ten billion as many photons as nucleons in the universe; and that ratio appears to have remained the same from close to the beginning of time to the present day.) The spreading of disturbances from the central mass created new space (in the instant after the disturbance) and that creation is still proceeding. Hence from Big Bang to the Expanding Universe.

I now have nothing worth saying about the further evolution of the universe nor how, if ever, it might end.

* * *

Now, at the end of this book, I must, for the last time, address the question: How well has it met the conditions for intelligibility stated in the first chapter? Earlier I argued for the reasonableness of my premises and the specificity of the formalism generated by them. I reaffirm those claims. I reassert also my claims to the depth and simplicity of my premises and of the formalism. As to causal connection, I have tried to show — how successfully is for the reader to judge — that the phenomena I have sought to explain connect, by entailment and therefore of necessity, in the models constructed in terms of the formalism, and I have urged that, through such connections, the paradoxes and mysteries that constitute the crisis of intelligibility in present-day physics can be resolved.

There remain adequacy and the complex of qualities in a theory or conceptual framework that give it Beauty. The models and explanation presented in this chapter, crude and fragmentary and petering out at the end, can hardly lay claim to adequacy. At best, they provide intimations of what is possible for those more capable than myself of fulfilling the purpose of this work. Only such persons can realize the Beauty of which I had an unattainable vision.